Urban Presences

Nio Architecten I Complete Works 2000–2011

目录

06 令人不安的现状
　　景观活力 | 萌动的领域

14 独眼巨人
24 力量之花
34 来自九的城市
36 地面控制
38 午夜石油
42 玫瑰呀玫瑰
44 推销马力诺
46 利物浦螺纹
50 高火山
52 波涛和旋风
62 柠檬花园

66 设计方法：对Maurice Nio作品的注释
　　特性 | 令人不安的现状

72 巴黎野玫瑰
84 屏息的爱
86 共存体
90 大卫和绿巨人
100 绿色小妖
104 绿巨人
112 感知浪潮
120 零点
124 野兰花

128 虚构的建筑
　　超越建筑 | 如何创建令人惊奇的基础设施

132 惊奇的鲸颚
146 水瓶座
156 我不是天使
162 亲吻新娘
164 安康鱼和水狼
168 月亮骑士
174 阴影下的祈祷
180 邪恶碰触
188 X-人

194 都市反思 | 城市新思路

196 黑盒子
202 第十一座住宅
210 天堂与地狱
218 大开口住宅
222 新领域
226 两个人的天堂
236 科学直觉
240 双面人
252 虫洞

256 垃圾焚烧场，荷兰特温特
　　无法言说 | 无法用语言表达的创建

262 蝴蝶夫人
266 暗物质
268 吃掉它！
270 热情的皇帝
278 蛇形空间
280 等到天黑
282 水印

284 建筑师简介

世界著名建筑设计事务所

NIO ARCHITECTEN
建筑事务所作品集

（荷）卡塔（Carta.S.） 编

江苏科学技术出版社

图书在版编目（CIP）数据

世界著名建筑设计事务所：NIO architecten /
（荷）卡塔（Carta.S.），蓝青编. -- 南京：江苏科学
技术出版社，2014.4
ISBN 978-7-5537-2811-7

Ⅰ. ①世… Ⅱ. ①卡… ②蓝… Ⅲ. ①建筑设计－作
品集－世界 Ⅳ. ① TU206

中国版本图书馆 CIP 数据核字（2014）第 011474 号

世界著名建筑设计事务所:NIO ARCHITECTEN

编　　　者	（荷）卡塔（Carta.S.）　蓝青
项 目 策 划	凤凰空间
责 任 编 辑	刘屹立
出 版 发 行	凤凰出版传媒股份有限公司 江苏科学技术出版社
出版社地址	南京市湖南路1号A楼，邮编：210009
出版社网址	http://www.pspress.cn
总　经　销	天津凤凰空间文化传媒有限公司
总经销网址	http://www.ifengspace.cn
经　　　销	全国新华书店
印　　　刷	北京建宏印刷有限公司
开　　　本	787 mm×1 120 mm　1/16
印　　　张	17.5
字　　　数	144 000
版　　　次	2014年4月第1版
印　　　次	2014年4月第1次印刷
标 准 书 号	ISBN 978-7-5537-2811-7
定　　　价	258.00元

图书如有印装质量问题，可随时向销售部调换（电话：022-87893668）。

Content

06 Silvio Carta | **Disquieting Presences**
Landscape Vitality | Spring the Territory

14 The Cyclops
24 Flower Power
34 From the Cities of Nine
36 Ground Control
38 Midnight Oil
42 A Rose is a Rose is a Rose
44 IL Sale Marino
46 The Thread of Liverpool
50 Volcano High
52 The Wave and the Whirlwind
62 IL Giardino di Limoni

66 Stefano Milani | **Modi Operandi. Notes on the Work of Maurice Nio**
Identities | Disquieting Presences

72 Betty Blue
84 Breathless
86 Coexistence
90 David and the Hulk
100 The Green Goblin
104 The Hulk
112 Sensing the Waves
120 Point Zero
124 Wild Orchid

128 Hans Ibelings | **Building Fictions**
Beyond Architecture | How to Create Amazing Infrastructure

132 The Amazing Whale Jaw
146 The Aquarians
156 I'm No Angel
162 Kiss the Bride
164 The Monkfish and the Waterwolf
168 Moon Knight
174 Prayer of Shadow Protection
180 Touch of Evil
188 X-Men

194 **Rethink Urban** | New Ways to Think about the City

196 Black Mothafucka
202 The Eleventh House
210 Heaven and Hell
218 House Wide Shut
222 New Territories
226 Paradise for Two
236 Scientia Intuitiva
240 TwoFace
252 WWIIOrmhole

256 Bart Lootsma | **Waste Incinerator, Twente, the Netherlands**
Unspeakable | Shaping Things that Cannot Be Told

262 Madame Butterfly
266 Dark Matter
268 Eat This
270 The Fire Emperor
278 Snake Space
280 Wait until Dark
282 Watermark

284 Maurice Nio + Joan Almekinders | Curriculum Vitae

Disquieting Presences
Maurice Nio, Urban Presences
Silvio Carta
Mar 2011

令人不安的现状
Maurice Nio, 都市存在

Silvio Carta
Silvio Carta Ph.D., Doctor Europaeus, architect and critic based in Rotterdam. Lives and works among The Netherlands, Spain and Italy. Finalist in the presS/Tmagazine competition for Young Critics 2007 and 2009 edition. Awarded in 2009 for the best Italian critic essay about new trends in architecture – unpublished category by the presS/Tmagazine critics competition. He regularly writes about architecture in the Netherlands, Italy, Korea and abroad in a diverse set of architecture magazines, newspaper and other media with reviews and critical essays. His writings have appeared in Antithesi, Newitalianblood, PresS/Tletter, Arch'it, A10 new European architecture, Mark Another Architecture, Frame, C3-Korea, Bouwelt. Since 2008 he is editor-at-large for C3-Korea magazine and books. He is currently researcher at the Faculty of Architecture - Cagliari (IT) and guest researcher at the Faculty of Architecture TUDelft (NL). He is involved in academic activities like inter-faculty researches, studios, seminars, workshops and lectures about contemporary architecture in different schools of architecture. He is currently carrying out a research about the role of criticism in architectural design. In 2008 he founded in Rotterdam the collective of critics The Critical Agency | Europe. In January 2011 he founded the online magazine TheArchHive, a Critical Archive of Architecture.

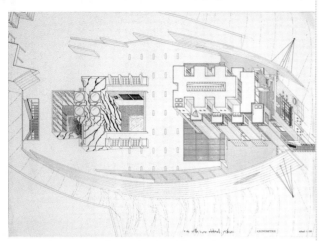

Nio's final thesis work 1- "The Villa for Michael Jackson"

Nio's final thesis work 2- "The Villa for Michael Jackson"

A Generic Panorama

Paul Virilio was one of the first intellectuals to depict the change of the urban condition from a planned system to a deregulated one[1]. In 1989, he described the contemporary city as "overexposed"[2] by reflecting on the configuration and meaning of the boundaries in the cities of the time[3]. The seminal (and various) cultural works of the intellectuals of so-called postmodernity by the end of the nineties had brought such a level of doubt concerning views of the world that society had become ready to look at the world as something continuously changing. Because reality was uncertain, the relationship between reality and its manifestations was considered a spiky question, difficult and dubious in the approaching.

From a cultural point of view, society was ready to welcome, or at least tolerate, a certain amount of doubt in terms of looking at the world. The modern principle of a world "conceivable as something objective" was definitely over and almost abandoned. It is worth noting that in 1989[4], it was relatively easy for young architects to feel comfortable in a world composed of the doubts, contradictions and complexities of the society.

In the same period, a distinctive sort of consciousness was emerging in various parts of the world. Passing through the reflection of several intellectuals such as—amongst others—British theorist Terry Eagleton, cultural considerations of the post-post modern condition were framed by a neologism coined by the French anthropologist Marc Augé. [5] The lucky expression Non-Lieux (1992) ended up being employed to indicate those places burgeoning from the supermodern society which were afterwards dismissed as generic. In his essay "Generic City,"[6] Koolhaas introduced that term in 1995 as a description of a contemporary city that Tschumi had appraised as "deregulated"[7]. Generic denotes that "general urban condition" which is "happening everywhere" and whose "very characterlessness provides the best context for living."[8] This particular adjective assumed a fundamental role in forthcoming architectural productions and theories. "Generic" are those places in which the (super)modern person gets lost[9] or has the feeling of being nowhere.[10]

In the Netherlands some years later, Dutch critic Hans Ibelings, in his essay "Supermodernism" (1998),[11] drew an outline of the architecture of supermodernity, an approach to architecture featuring a "sensitivity to the neutral, the undefined, the implicit; qualities that are not confined to architectural substance but also find powerful expression in a new spatial sensibility."[12]

Maurice Nio

Having clarified this generic scenario, we can consider where and how to place—although with some careful precautions—the figure of Maurice Nio.

Nio studied at the Faculty of Architecture at Delft University of Technology for about ten years. During this period, he cast doubts on what architecture really was. "I actually did not want to be an architect after my first year at the TUDelft," he says. "I wanted to be a film director, but stayed at the university. The first year, I started as a regular student, following the prescribed program, but in the second year it became clear that architecture was merely about… designing buildings. The extra was maybe some kind of social sauce to distract one from the fact that it was an empty discipline. Nothing exciting."[13]

The diffused perception of architecture as something truly ineffable soon became for Nio a primary reason for finding his own way in the architectural field. "It was impossible to 'learn' architecture, not in the university, not from good architects outside of the university." For this reason, Nio sought cultural references in other cultural fields: "We as students had to develop our own program. The official architecture program took five years. We (some friends and I) created an unofficial program which also lasted five years, consisting in understanding philosophy (reading Foucault, Deleuze, Baudrillard, all exciting French writers, but also German writers like Nietzsche, Benjamin, Kafka, Canetti), translating books like Les Strategies Fatales of Baudrillard, experimenting with time-related media-like video. "

Maurice Nio's reaction to the difficulties of understanding a continuously changing reality and producing architecture with a unique system of values would translate into an attempt to enlarge the range of elements in architecture. "Before architecture can exist, it can be understood" he explains, "there should be first the understanding of culture and the position of architecture in culture. During my ten years of study, the most problematic thing was to bring theory and practice (architecture) together."

Nio's final project at TUDelft represents an important landmark in his work. As he was initially more interested in filming rather than building, his first proposal for the graduation project was "to design the setting for a movie, but my teacher said that it was not possible: it should be a building with construction and details," he explains. "So I decided to make a design for the villa of Michael Jackson (which is almost a setting for a movie)."

After his degree, Nio "saw a chance to be an architect in a 'liberated' way, where theory and architecture (even on a commercial, social, functional level) are not opposed but intertwined in an ambiguous way." As a consequence, he became involved in the cultural atmosphere of the Netherlands.[14] Nio is locally known for his translation of Baudrillard's book Amerique. What Nio reports from this period is a reflection on the coherence behind the architect's work. "An enormous weight fell off my shoulders—the weight of coherence… the desire for coherence is anywhere and always tangible, especially within the circle of writers, artists, architects, etc."[15] Within the construction of the picture of a fragmented and elusive reality which other intellectuals were working on, Nio speculated about consistency in the work of the architect. "Where most of the architects have managed to build a beautiful coherent style, with me every project seems to erase the memory of the image of the previous one".[16] Finally Nio reaches a conclusion which can be seen as a guideline he would follow in his forthcoming works: "I understood that coherence could not be found in style, form and the explicit, but in method, structure and the implicit."[17]

Nio's final thesis work 3- "The Villa for Michael Jackson"

Visions and Technical Concerns
If method, structure and the implicit are key points for Nio in design, it is understandable that his architectural productions will not present similarities in terms of language or elements.[18] The coherence —if we wish to label it such—or the red thread running through Nio's work, has to be found in the intention to confer on each project its own identity. Whereas Nio's background picture of the world is of such a fast and changeable (unpredictable) reality as to make cities generic and buildings blurred in a new panorama each day, he reacts by staking his position in all that. To try to confer a univocal identity upon each project means to pop the building out of a generic background. Maurice Nio's answer to Koolhaas' generic city is a city dotted by presences. From The Hulk to the Betty Blue, each of his projects has its own name and—if you like—its own story to tell.[19] Each project starts with a fascination, an image or a sequence which leads the project to its final result.[20] The method implied by Nio consists of two parallel paths in which the design proceeds. On the one hand, there is the initial image, the "beginning fascination" which gives the first sparkle and which appears several times during the process as a "moment of truth" (key) to verify the coherence (used literally) of all the developments; and the initial fascination is eventually used to confront the found outcome. The name of the project closes the circle by compressing all the phases within a meaning, suggested by a name. The name is of course not coined by the architect: he needs his project to refer to a story which already exists as embodied in its title. The name thus stands for the meaning that Nio wants to confer upon the project. Each name is thus chosen for its evocative power.[21]

The second direction the projects follow stems from technical concerns. On the one hand, if the fascination issue leads and directs the project from scratch to the end, its feasibility, on the other hand, makes the project buildable and thus possible. Nio's team pays a great deal of attention to technical aspects in every project. The structure and the air conditioning have to work "as a project within a project," as "a machine." Nio knows that a visionary project can be trusted (by clients, tenants, investors, city halls…) and thus rendered only if it comes with a clear and precise engineeristic (pragmatic) system of solutions. If the vision pushes the design over the conventional boundaries, the technical system brings it back to the world of city regulations, budgets and physical materials, and thus makes it possible. The more a vision is needed to explore a mutating reality, the more a working system of (technical) solutions makes it happen. The two main directions run at the same speed in Maurice Nio's office, although sometimes one may outstrip the other. In The Fire Emperor—the competition for Rotterdam's new market hall—for instance, the images Nio creates describe a world deformed by the presence of food in all its declinations and possible forms. He explains:
"The supply of food and drink are often a reason to go out of the house, into the city. Everybody has to eat every day. For some, this daily routine of buying, cooking and eating has even become a ritual. All these routines and actualities we want to bring closely together, so the intimacy and cultural richness of the food become clearer, more visible. That is where the public and the private life can tolerate each other wonderfully."

As a counterpart, several of Nio's projects realized in leftover spaces (often in correspondence with a flyover) can be called into consideration. In these examples, the intention to recover and reuse a leftover space, say a leftover condition created amongst the main urban and infrastructural elements—streets, train ways, parking and so on—is clearly visible. In such projects, a spatial and pragmatic need demands a vision able to revitalize particular urban spots. The usefulness and the requirements lead the project, whereas—somehow—the vision aspect makes it possible and desirable for investors, municipalities and clients in general.

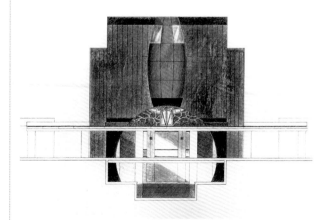

Nio's final thesis work 4- "The Villa for Michael Jackson"

Seeking an Identity
It should be emphasized, however, that the two main strategies (or tools) Nio employs in his work have the construction of identity as a final goal. This aspect is the result of the combination of three factors. The first

is—as we have seen—related to his education and cultural background—his studies of the cultural atmosphere in which he started to ask himself which direction to follow—an intense and complex period in which Paul Virilio, Rem Koolhaas, Bernard Tschumi, Jean Baudrillard, Noam Chomsky, Jacques Derrida were just a small part of a wide corpus of intellectuals of reference.

The second aspect stems from the position Nio decided to take within that cultural framework. Feeling, and subsequently taking as a starting point, the consciousness of an elusive, anonymous architecture, he decided to set the trend on its head by challenging it with an architecture featuring an extreme identity.

The third point is intimately related to Nio's peculiar character and his personality. It seems to be among his personal intentions to focus on contradictions within reality. Contradictions reveal the fallibility of the interpretative tools we use in understanding reality, in full accord with the conditions of postmodern culture.[22] The importance of revealing the contradictions was something that had already been declared by Robert Venturi in 1966.[23] However, Nio seems interested in investigating those contradictions and using that investigation to react to generic-ness. Working with contradictions means undermining certainty. In other words, it means breaking the quietness of a clearly understandable and measured scenario. The presences Nio places in the city are thus disquieting. They need to present a certain amount of oddness to work properly.

Considering the type of projects Nio has dealt with can also assist in this understanding. The list ranges from incinerators to movable bridges, from flyovers to sound-barrier housings. In terms of visions, he combines animals (the appearance of beasts) such as wolves (The Monkfish and the Waterwolf), birds (the competition for Oristano), whales (the Amazing Whale Jaw) or flora, as with the White Orchid. He even involves fantasy forms like the half moon in the extension of the Museo Pecci in Prato (Italy) or the X-men. In any case, he builds perfectly working machines out of his visionary forms. The Garzetta bird for Oristano becomes a remarkably elegant urban object in which the wings are the car ramps to access the first floor of the car park and the beak (the tower of the bus terminal) is an air duct. Moreover, the oblong horn placed in the middle of the semi-circular sharp volume of the Pecci extension works as a structural element. This last example is emblematic. The weirdly protruding horn shape may appear gratuitous at first sight, but it confers on the new Pecci Museum a globally recognizable (yet elegant) identity while at the same time addressing technical concerns: "It is an antenna, horn or sensor. It scans the waves of the future. And in a practical and functional sense, it is the stabilizer when an earthquake will occur," Nio explains. The horn is representative of the design method of Maurice Nio. In a sense, it is the essence of his architecture.

A further aspect should be highlighted at this point. The plants, animals, and shapes Nio uses in his design are chosen in an ambiguous light. The animals are wolves, praying mantises, reluctant or ferocious species caught in extremely elegant or proud poses. The clearest example of this tendency is the stuffed pelican whose beak is impending a few centimeters from the meeting table at his office in Rotterdam. A meeting in his office starts with this bizarre yet elegant image from nature.

Shapes, originally basic shapes, during the design process are deformed, corrupted, disfigured—whereupon the elegant yet intriguing appears. The interest is heightened by the lack of clarity: a building suggests a significance through its shapes and name, but does not declare it openly. Stories in Nio's projects are hinted at, but not definitively explained. This difficulty makes the stories unclear or—better—quasi-clear, which is the quality that inspires curiosity and attention. His projects are allusive, indeterminate and therefore ambiguous and disquieting at the same time, especially placed as they are against the background of a generic urban panorama.

Poggiomarino - "Basic instructions before designing the train station Vesuvio Est."

What is the Value of Maurice Nio's Work in Contemporary Architecture?

The question which necessarily closes a presentation of Nio's work concerns the results he has thus far achieved as he pursues his conception of architecture. Having clarified the premises of his research as executed through his several projects, we may speculate as to what he achieved in his buildings and other built identities—but to venture a final answer is premature. In 2011, Nio has a great many projects to work on and—as we have seen—his approach to design is under continuous development.

The beginnings of an answer, however, can be articulated as follows. Some architects realize extremely precise buildings, highly technological or intelligent, sustainable and almost self-sufficient, buildings with the aim of representation or honoring cultural heritage, or even buildings that are experiments in new ways of living. Amid all this, Nio does not focus on one building as a part of his personal way of conceiving of architecture.[24] Rather, he has built over the years a family of monsters based on flora, fauna and inanimate objects—an entire world of presences created to disturb tranquility and constantly raise questions and doubts about the way we see reality every day, because our contemporary reality is doubtful. Will these disquieting presences finally emerge from the generic panorama?

Captions:
[1] Cf. Bernard Tschumi, "De-, Dis-, Ex-," in B. Kruger and P. Mariani, (eds.), Remaking History (Seattle: Bay Press, 1989).
[2] "The urban has lost its form, with the sole exception of the form image without dimension, the point or punctum which is everywhere, while the measurable length is nowhere. In the manner of the nodal or Pascalian mode, this center which rejects all circumference and even the very concept of periphery is the uncertainty principle applied to the world geomorphological continuum." Paul Virilio, "The Overexposed City" ("La ville surexposée"), from L'espace Critique (Paris: Christian Bourgeois, 1984); translated by Astrid Hustvedt in Zone 1–2 (New York: Urzone, 1986).
[3] For further readings, see Paul Virilio, L'espace critique, op. cit and Bernard Tschumi, "De-, Dis-, Ex-," op. cit.
[4] The date corresponds to "De-, Dis-, Ex-," Bernard Tschumi's essay, but it is also quite important to remember that on November 9 of the same year

the Berlin Wall officially fell, initiating a re-configuration of the world's arrangements and initiating a new balance among nations.

5 Marc Augé, Non-Lieux: Introduction à Une Anthropologie de la Surmodernité, Le Seuil, 1992.
6 "The Generic City" is the concluding chapter of the book S, M, L, XL by Rem Koolhaas and Bruce Mau, published by Monacelli Press in 1995 in New York.
7 Bernard Tschumi, "De-, Dis-, Ex-," op.cit.
8 Rem Koolhaas, "From Bauhaus to Koolhaas," interview in Wired 4.07, July 1996.
9 Cf. the film Crash, 2004, directed by Paul Haggis, and Concrete Island, a 1974 English novel by J. G. Ballard.
10 We might say "elsewhere": cf. Marc Augé, "Near and Elsewhere," in Marc Augé, Non-Lieux: Introduction à Une Anthropologie de la Surmodernité, (Le Seuil, 1992), English translation: John Howe, (Non-Places: Introduction to an Anthropology of Supermodernity), Verso, London, 1995.
11 Hans Ibelings, Supermodernism: Architecture in the Age of Globalization, Nai Publishers, Rotterdam, 1998.
12 Ibidem.
13 All quotes from Maurice Nio in this text, except those from the Peak publication, are from interviews with the author between December 2010 and January 2011.
14 During the eighties the Netherlands recognized the necessity of opening its architecture to the "cultural component of architecture." Cf. Bart Lootsma, "SuperDutch Afterthoughts," in Post.Rotterdam, 010 Publisher, Rotterdam, 2001.
15 NIO Architecten: 02 Design Peak, Equal Books, Korea, 2010.
16 Ibidem.
17 Ibidem.
18 The work of several architects is recognizable for the presence of key elements or features that are part of the architect's vocabulary. One example is the use in the oeuvre of Aldo Rossi of certain shapes which he inherited from his childhood.
19 The relationship between the name of the film and the specific project can be found in Nio's explanation: "The titles of the projects are not always titles of films. Sometimes we dream a film. In a way all projects are stories, as all films are dealing with a narrative line." (From an interview with the author in January 2011).
20 Maurice Nio explains concerning the genesis of a project that "there is not really a start. The projects were already started. We just jumped in, on a moving train. If we are lucky we find an image we can hold on to, but most of the time we are puzzled and lost in the process. If we are lucky we understand the project when it is built, but sometimes we still cannot grasp its meaning. We can communicate with clients and contractors (of course) but we do that in the language which everybody understands." (From an interview with the author in January 2011)
21 The Incredible Hulk, named for the Marvel Comics character, suggests the features of the Waste Incinerator in Twente (NL): an enormous monster that performs superhuman feats like pulverizing and consuming garbage. ("Hulk crush!" the character says in the Hollywood movie.) The monster is—obviously—lit by strong green lights.
22 Although PoMo has early origins, the cultural scenario described in the first part of this text can be seen as one of the consequences of the PoMo period.
23 The book Complexity and Contradiction in Architecture was published by The Museum of Modern Art Press in New York in 1966.
24 To do so is precisely what Nio denied by calling it "style"; cf. the interview in 02 Design Peak, Equal Books, op. cit.

Note: from above, the images of Nio's final thesis work and Poggiomarino are provided by Maurice Nio.

通属概论

Paul Virilio是最早关注城市从规划系统解放出来之后的环境变化的学者。早在1989年，他就提出当代城市因城市边界线的形状和意义而被"过度曝光"的观点。20世纪90年代末，许多有影响力的后现代文学作品都提出了"社会已经准备好接受世界的不间断变迁"的观点。现实及其表象之间的关系由于前者的不确定性而变得尖锐，从而备受质疑。

从文化的角度来看，社会对此持欢迎态度，至少能够以容忍和部分怀疑的态度看待这个世界。"相信客观世界"的现代原则几乎被人们摒弃。值得注意的是，在1989年，年轻的建筑师相对容易适应充满怀疑、矛盾和复杂性的社会。

与此同时，世界各地的学者提出了各种不同的见解。英国理论家Terry Eagleton、法国人类学家Marc Augé通过新的词汇建构了后现代文化之后的文化因素。Non-Lieux（1992年）指出超现实社会中迅速发展起来的各种场所普遍走向消解。Rem Koolhaas（1995年）在他的文章《通属城市》中介绍了这个用来描述当代城市的概念。Bernard Tschumi曾用"放松管制"来形容该术语。"通属"指的是"无处不在"的"城市的总体条件""以毫无特征的方式提供了最好的生活环境"。"通属"这个形容词假定了未来建筑实践和理论的基础性作用。在一个"通属"的超现实社会中，人们会迷失自我，不知身在何处。

数年后，荷兰评论家Hans Ibelings在他的文章《超现实主义》（1998年）中描绘了超现实主义建筑的轮廓，这种建筑"对中性、不确定性和模糊性非常敏感；建筑的特质不会受限于建筑物本身，它以一种全新的空间敏感性找到了有力的空间表达形式。"

Maurice Nio

在解释完"通属"这一概念后，我们可以通过Maurice Nio的经历得知这一概念是如何变成现实的。

Nio曾在代尔夫特理工大学建筑学院学习了十年。在此期间，他对建筑的本质产生了怀疑。"在进行完第一学年的课程后，我当时并不想成为建筑师。"他说，"我想成为一名电影导演，但最终还是留在了学校。第一年我仅仅完成了规定课程，但在第二年我觉得建筑的概念变得清晰起来，它们是凝结设计师奇思妙想的房子。那些额外的社会因素会分散人们的注意力，但建筑本质上是一个空洞的学科，没有什么激动人心的地方。"

建筑的发散思维确实是不可言喻的，Nio很快就找到自己的建筑之路。由于"建筑是无法从大学和优秀建筑师那里'习得'的"，Nio开始从其他文化领域寻求答案：
"当学生的时候，我们要学会发展自己的方案。正式的建筑方案需要用五年的时间完成，我和几个朋友做非正式的方案也用了五年，我们看哲学书，读了法国作家Foucault、Deleuze、Baudrillard的作品，还有Nietzsche、Benjamin、Kafka和Canetti等德国作家的著作，翻译了Baudrillard的《致命的策略》，进行视频制作这类与时间有关的媒体试验。"

不断变化的现实理解起来困难重重，用与众不同的价值体系思考建筑也实属不易，Maurice Nio尝试通过扩大建筑元素的范围来解决这些问题。他解释道："在建筑盖起来以前，我们就可以理解它们。首先要理解文化和建筑在社会文化中的地位。在我十多年的研究中，最棘手的事情就是将理论与实践（建筑）结合在一起。"

Nio的在代尔夫特理工大学的毕业设计是他个人作品的一个重要里程碑。当年他对电影的兴趣超过了建筑，他曾设想以电影场景作为毕业设计。但这一提议被老师拒绝了：必须设计一座有结构、有细部的建筑。他解释说："所以，我决定为迈克尔·杰克逊设计一座别墅（这几乎是一个电影场景）。"

毕业之后，Nio"得到了以'解放'的方式做建筑的机会，理论和实践（即使在商业、社会和功能的层面上）本不应该对立，而应该以模糊的方式交织在一起"。因此，他加入了荷兰的文化界。Nio因翻译了Jean Baudrillard的《美》而声名鹊起。他在自己的一系列作品中贯穿了同一条主线："一个重担压在了我的肩膀上，我要在作品中体现一致性……这一要求对于作家、艺术家和建筑师更为明确。"别的学者都在研究纷繁复杂且不断变化的现实世界，Nio却一直在思考如何在建筑作品中保持一致性。"大部分建筑师已成功地建立起自己的风格，而我的每一个项目似乎都要抹去前一个作品的印记"。最终，Nio形成了自己在作品创作中的准则，这些准则会体现在其下一个作品上："作品的一致性不应体现在风格、形式等明确的表象上，而应体现在设计方法、建筑结构和某些不言自明的东西上。"

想象力和技术

如果设计方法、建筑结构和某些不言自明的东西是理解Nio作品的关键，那么他的建筑作品在设计元素或语言上就没有相似之处。Nio作品中的连贯性，或者说贯穿始终的主线赋予每一个建筑以个性。然而，快速变化的世界充满了不可预知性，新建筑的形象模糊在普通城市中，Nio努力划清自己与它们的界限。赋予每个建筑以明确的身份就要让它们从普通城市的背景中跳出来。Nio用散点式的城市景观回应Rem Koolhaas的"通属城市"。从废弃的船到巴黎野玫瑰，他的每个项目都有自己的名字和故事。他的建筑要么开始于一个传说，要么有一个建筑场景贯穿始终。Nio的设计作品立意与技术并重。最初的建筑场景给他灵感，并在设计过程中数次出现，作为"建筑真相或关键时刻"见证了所有发展阶段的一致性并导致最终的设计结果。项目的名称用简练的词语道出了建筑的意义，这个名字当然不是建筑师凭空创造出来的：他会参考能够体现项目主题的故事，并使用故事的名字。这个名字就代表着Nio想要赋予项目的意义，它们会唤起建筑内在的力量。

除此之外，项目还必须遵循技术原则。如果说神话故事贯穿于方案设计的全过程，那么方案的可行性就使得项目的建成成为可能。Nio的团队十分关注每个项目的技术层面，建筑的结构和空调系统是"项目中的子工程"，是为建筑服务的"机器"。Nio深知，一个建筑方案可以得到客户、租户、投资者和政府的认可，但只有通过清晰、明确的实施方案才能将其变为现实。如果说想象力使方案设计突破了传统的束缚，那么技术系统就把方案带回了城市法规、预算和建筑材料的世界，并使之成为现实。想象力创造的世界与现实差距越大，将其实现所需要的技术支持就越多。在Nio的工作室里，这两条主线不一定同步发展，但却同等重要。例如在鹿特丹的新市场方案竞赛中，Nio提出了热情的皇帝的设计方案，他用建筑形体的倾斜和其他形式表现了因食品现状而变化的世界。他解释说：

"人们每天都要吃饭，食品和饮料的供应应该走出建筑而进入城市。对一些人来说，购买、烹饪和品尝食物已经成为一种仪式。所有这些过程都应该紧密结合起来，丰富的食品文化会因此而变得更加清晰。这样一来，公共生活和私人生活也可以巧妙地结合起来。"

Nio有相当一部分作品位于被人们遗忘的角落（通常与高架公路相关），他对这些建筑有很深的思考。在这些例子中，我们可以很明显地感受到设计师试图恢复和重新使用那些剩余的空间，例如，街道、铁轨、停车场等城市中的基础设施。在这些项目中，功能上的需要要求建筑师用想象力恢复特定地段的活力，而那些想象同时要具有可操作性以说服投资者、政府当局和委托人。

寻找建筑的个性

应该强调的是，Nio在工作中使用的这两种策略（或工具）都是为了创造建筑的个性。三个因素导致了这个目标的提出。首先是他本人的教育和文化背景，他在对文化氛围的研究中寻找自己的方向。在一段紧张且复杂的时期中，Nio吸收了很多人的思想，包括Paul Virilio、Rem Koolhaas、Bernard Tschumi、Jean Baudrillard、Noam Chomsky和Jacques Derrida等。

第二个因素源于Nio在文化中所汲取的东西。一般的设计过程通常是在获得灵感后立即选取一个切入点，在一种难以捉摸的感觉下产生一幢毫无特色的建筑。Nio则运用独具特色的建筑挑战这种创作过程。

第三个因素与Nio独特的性格特征密切相关。它似乎天生关注现实中的矛盾。这些矛盾揭示了我们对现实的错误解读，这些误解与后现代文化的实际情况完全吻合。罗伯特文·丘里已经

在1966年揭示了一些矛盾，然而Nio似乎想用这些调查反击"普遍性"，换句话说，他想打破已经被多数人理解和接受的事实。Nio眼中的城市现状令人担忧，他认为城市中需要引入一些异质物，人们才能正常工作。

Nio所做的项目类型也有助于我们理解这一点。他的项目包括垃圾焚烧场、可移动桥梁、公路立交桥和隔声建筑。在建筑外观上，他将动植物的外表与建筑结合起来，例如，贪婪的动物（安康鱼和Waterwolf）、鸟类（Oristano设计竞赛）、鲸（令人惊异的鲸鱼颚）或白兰花。他甚至在意大利普拉托的Museo Pecci扩建项目中使用了半月形和X战警的形式。在任何情况下，他都能从建筑外观出发设计完美的机器。Oristano设计竞赛中的白鹭方案优雅而引人注目：白鹭的翅膀是通往一层停车场的汽车坡道，鸟喙（塔的终端）则是一个风道。长方形的喇叭作为结构的一部分与Pecci延伸部分的半圆形的体量相连。这个例子具有象征意义，那些古怪的形状乍看起来似乎没什么道理，但它赋予了Museo Pecci整个建筑以优雅的个性，同时解决了技术问题。Nio解释道："这是天线、喇叭和传感器，它会扫描到未来的波。在地震发生的时候，这里还可当做避难所。"这个喇叭代表了Maurice Nio的设计方法，是整幢建筑的精髓所在。

另一个应该强调的问题是Nio在设计中选用的动植物形象都较为抽象。这些狼和螳螂的形象很优雅，完全没有捕食者的凶残。最好的例证是鹿特丹的一座办公楼，一只饱腹鹈鹕的嘴部距离办公桌仅有几厘米的距离，在这间办公室里进行的会议要从这个奇怪却不失优雅的形象开始。

在设计过程中，最初的几何形经过变形，最终优雅而又耐人寻味。建筑的趣味源自其不确定性：建筑的形状和名称暗示了它的重要性，但建筑本身并没有公开宣扬这一点，Nio的建筑就是如此。这使建筑的故事不那么明确，只能靠高质量的空间才能引发人们的好奇心和注意力。他的建筑具有间接性和不确定性，特别是将它们置于通用城市的背景中时，更会令人感到含糊不清和不安。

Maurice Nio的作品在当代建筑中的价值

我们用这个问题来结束对Nio作品的描述，转而谈一谈他对建筑观念的贡献。历数这些年来他曾经研究过和已实施的项目，我们认为他在创造建筑个性方面有一定的贡献，虽然这么说似乎还为时过早，因为2011年，Nio还有很多方案要实施，他的设计方法还处于不断发展中。

然而这个问题的答案并不简单。建筑师实现了一座具体建筑的设计方案，这些建筑要么是高智能建筑，要么是可持续建筑（在能源上可以自给自足），要么是对文化遗产的保护，有些甚至是实现新生活方式的实验性建筑。但Nio的建筑并不是为了实现自己的建筑构思，相反，他以植物、动物和无生命的物体为基础，制造了很多"怪物"来扰乱世界的安宁，质疑我们生活的现实世界，因为在他眼中，当代的现实是难以预料的。这些令人不安的现状是否会出现在普通场景中呢？

备注：中文翻译没有罗列原英文中部分内容的引用来源，如有阅读需要，请参考英文引用索引。

Landscape Vitality
Spring the Territory

景观活力
萌动的领域

The Cyclops
Flower Power
From the Cities of Nine
Ground Control
Midnight Oil
A Rose Is a Rose Is a Rose
IL Sale Marino
The Thread of Liverpool
Volcano High
The Wave and the Whirlwind
IL Giardino di Limoni

THE CYCLOPS

12 Soundbarrier Houses in Diependaal, Hilversum

Address: Charley Tooropstraat, Hilversum, Holland
Design: VHP s+a+l/NIO Architecten
Client: Slokker Vastgoed bv
Contractor: Slokker Bouwgroep bv
Structural engineer: Adviesbureau Steens
Design team: Remco Arnold, Eric Lucas, Maurice Nio, Jaakko var 't Spijker
Start design: 1997
Completion: 2001
Costs: euro 1,495,450.00

独眼巨人

希佛萨姆第潘达的12座隔音住宅

Photo credit: Andrew Thurlow, Rob Ponsen

The twelve houses form an integral part of a soundproof embankment along a secondary road from Diependaal, an exclusive residential district in the woods of Hilversum. Despite its inhospitable setting and the clients' initial scepticism, the futuristic, not to say fantastic, design of the houses attracted users who bought their houses of the drawing board, several years before delivery.

Without what is in effect a screen of houses, allowing the level of ambient noise to be brought down to acceptable limits, the rest of the urban design plan for Diependaal would never has seen the light of day. The project's congenital, altruistic handicap justified the atypical solutions eventually proposed by Nio and his associates in the face of the client's reservations. So for example the answer to the darkness at the back of the plot was to cantilever out the living room at the first floor level. Moreover, not everyone considers the lack of a garden to be a drawback, and a terrace can be a sales argument, even in suburbia.

In Hilversum, as in any fantastic setting, a house can be both a cavern and a giant, a term that could just as easily refer to the block of flats opposite. The residents are all explorers who are just passing through, threatened by poisonous rays and nostalgic for a place of peace and unspoiled nature that only a handful of neo-modernistic architects, obsessed by the American subculture and pessimistic prophecies, still dare to denounce publicly.

这12座住宅形成了隔音路堤的一部分，分布在从Diependaal延伸过来的二级公路沿线，这里也是Hilversum林区唯一的可居住区域。虽然基地周边的荒凉引发了客户最初的怀疑，但这个前卫的设计（这并不是称赞设计很出色）还是使他们在建筑交付使用数年前就为设计方案付款了。

只有将环境噪声水平降低到可以接受的限度，为建筑提供有效的隔声屏障，Diependaal的城市规划才能真正实现。Nio和他的同事利用该项目周边原有的音障，创造性地解决了客户的需求。例如，为了避免基地后面出现黑暗角落，一楼客厅向外伸出了悬臂结构。此外，客户也接受了没有花园的住宅，建筑的露台在郊区成为了卖点。

在Hilversum，面积大的住宅（与公寓相对）很受欢迎。居民都有居住小公寓的经历，受到过有毒射线的威胁。他们怀念宁静的住所和未受污染的自然条件。只有少数的新现代主义建筑师（他们大都信奉美国亚文化，对未来感到悲观）仍不敢公开谴责小型公寓住宅。

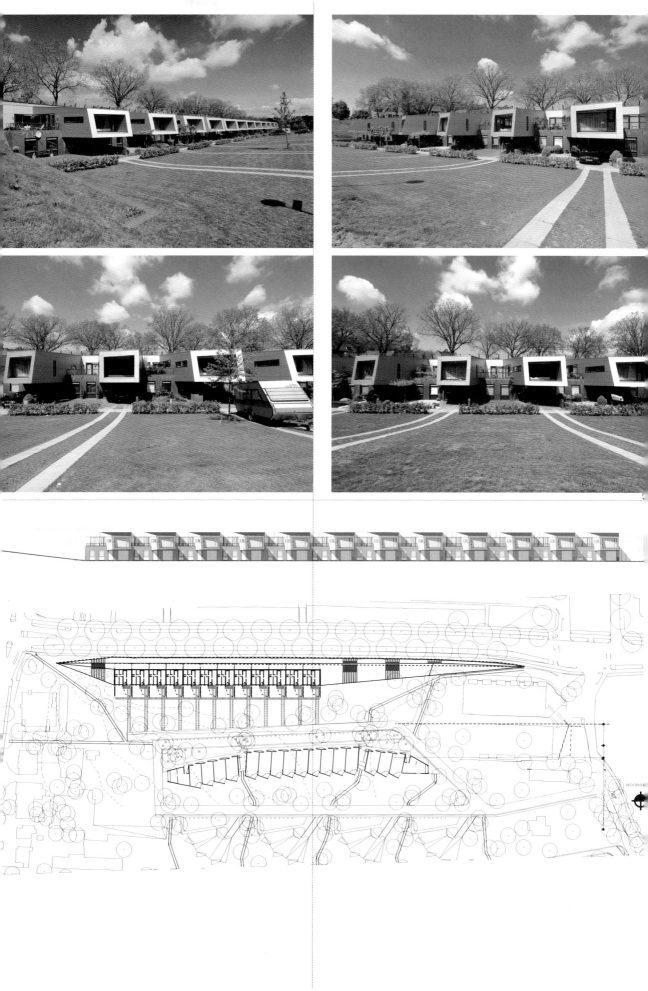

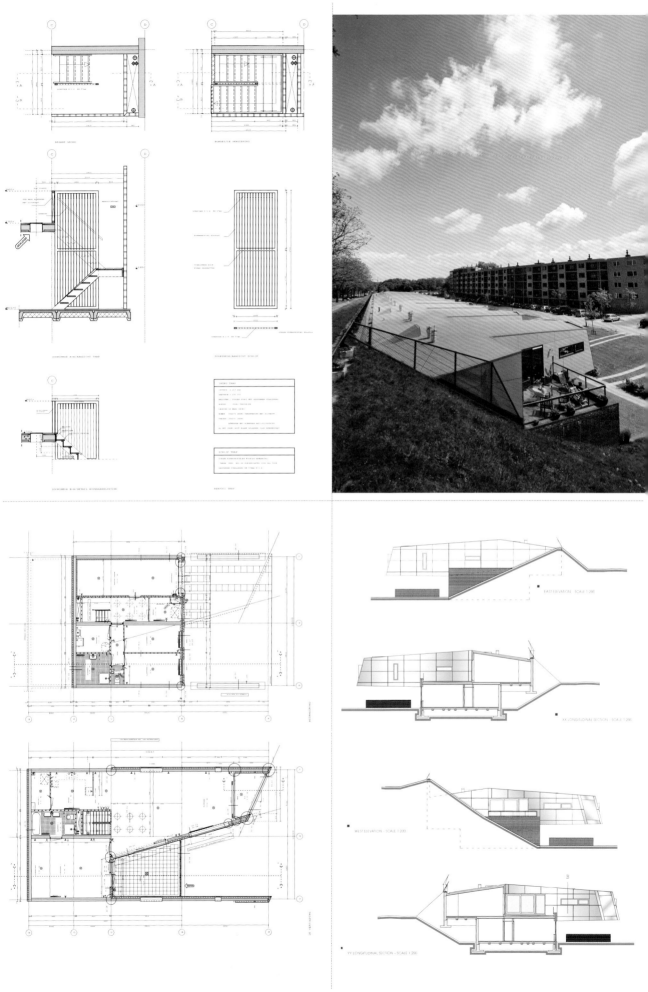

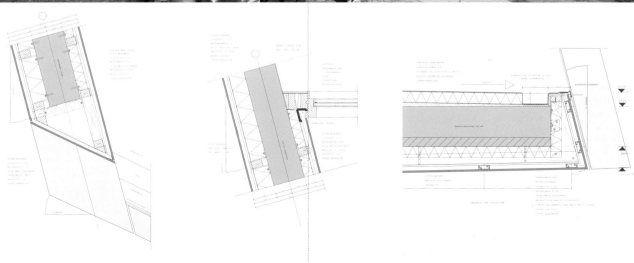

FLOWER POWER

Southern Tangential High Speed Bus Track (Kerntraject Zuidtangent)

Address: Schiphol/Haarlemmermeer/Haarlem, Holland
Design: VHP s+a+l with Dok industrial design
Project manager architecture: Maurice Nio
Project manager industrial design: Gijs Ockeloen
Project manager landscape architecture: Paul Kersten
Commissioning authority: Schiphol Project Consult
Building contractor: Cluster 4
Structural engineer: ABT & Holland Railconsult & DHV & IBZH
Design team: Hernando Arrázola Castillo, Riëtte Bosch, Henk Bultstra, Mirjam Gaijé, Paul Kersten, Maurits de Koning, Fevzi Köstüre, Rop van Loenhout, Maurice Nio, Gijs Ockeloen, Mariska van Oosterhout, Stephen Pasterkamp, Germaine Sanders, Jaakko van 't Spijker, Claudia Teriou, Gijs Wolffs
Take-off design: 1998
Completion: 2002
Building costs: euro 250,000,000

力量之花
南部高速公交线路支线

Photo credit: Hans Pattist, Go Designers

Think of Dutch public bus transport and what you see is minimal bus shelters: one small plastic bench and standard 30 by 30 paving stones. If you want to be polite you could call it "rational" or Calvinistic, but actually it is simply poor. No wonder no one takes the bus with pleasure. With the design of the Kerntraject Zuidtangent, we wanted to boost the image of public bus transport by reacting against this common picture. We wanted the steel to dance and the concrete to speak. We wanted to cross technique with flora, Schiphol with the Floriade, super service with supra-identity. This way the Kerntraject Zuidtangent could instantly be given a face. Passengers had to experience that they were making use of something special. The common image of the public transport-passenger, who is standing waiting in a cramped bus shelter, would get a counterpart with style: there is bus transport and there is the Zuidtangent!

Because the bus lane traverses various areas (residential areas, inner city areas, industrial areas, office locations, greenbelts and tunnels), we have conceived all landscape, urban development, architectural and civil facilities for the Zuidtangent as a continuous system to make the image of the Zuidtangent appear as powerful and as clear as possible. With recurring elements – such as coverings, fencings, viaducts, windscreens, sound barriers, tunnels – a system has been designed in which subtle variations are possible. All bus shelter facilities are black and white, apart from the alternating coloured glass roofs (and the bicycle tunnels in the Haarlemmermeer).

The 14 bus stops of the Kerntraject Zuidtangent are partly situated on works of art. Consequently, the concrete constructions can never ever be conceived of as neutral platforms. You simply cannot dissociate the baroque high-tech shelter facilities of the other much-needed civil interventions. Quite some time has been put into adapting the already-worked-out-but-not-yet-designed concrete works of art, in which the cross-section of the viaducts is optimised in order to get a sharper and smaller construction. Within all restrictions, a more elegant road surface profile has been designed. Underneath this profile, expressive solid supports predominate. Therefore, the whole of works of art stands out in a rather fierce way against the flowing coloured glass bus shelter-coverings.

The bended steel fencings, together with the thin prefab concrete panels form the finishing of the edges of the profile of the viaduct. The trick was to avoid the traditional beam/T-bar construction, or at least to make it invisible. This way, what is created in appearance is the desired continuous profile above which the red Zuidtangent-busses skim along at a high speed. The pillars that have been round off at the edges and that have been placed slantwise, literally hold up the bus lane. They are musclemen made of self-condensing concrete. Sometimes, as is the case with the viaduct of the Spaarne hospital, they actually do come out of the ground, as if they give the bus a little push for it to go even faster.

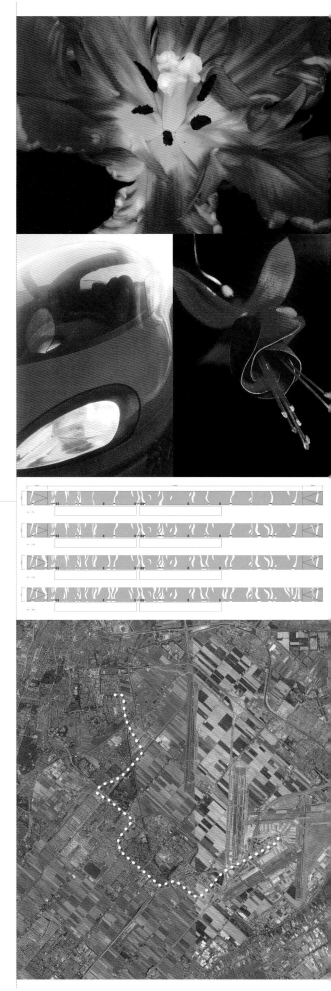

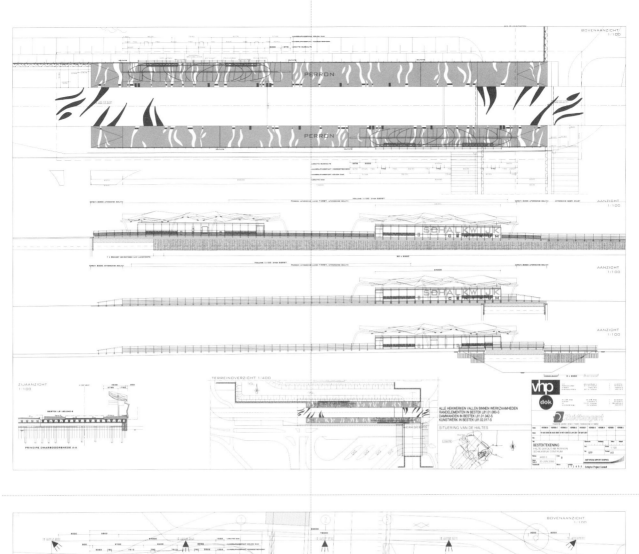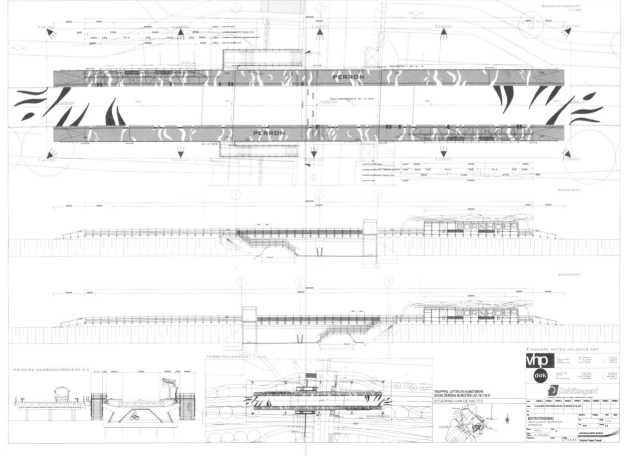

knooppunthalte

standaardhalte

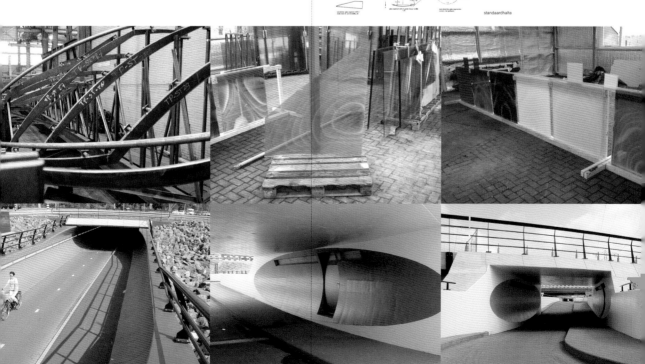

回想一下荷兰的公交站和你见过的最小的候车亭：一个塑料小条凳和标准的30厘米×30厘米的地砖。客气一点的话可以称之为"理性"或"加尔文主义"，但候车条件确实很差。难怪人们坐公交车时心情总是不愉快。我们想通过设计Kerntraject Zuidtangent来提升公共交通在人们心目中的形象。为了让钢铁舞蹈起来，让混凝土开口说话，我们想把技术与植物、Schiphol和Floriade、良好的服务和独特的个性结合在一起，建立Kerntraject Zuidtangent的新形象。乘客会切身感受到他们正在使用一种独特的东西。原先在拥挤的公交车站等车的乘客会得到全新的体验：这就是Zuidtangent的公交车站！

公交线贯穿了城市的不同区域（居住区、内城、工业区、办公区、绿地和隧道），我们将景观、城市发展、建筑及当地的民用设施结合起来，形成一个连续的系统，使得Zuidtangent的形象更加有力、清晰和明确。我们对一些重复出现的元素（如覆盖物、篱笆、高架桥、防风罩、声屏障、隧道等）进行系统化设计，在其中加入微妙的变化。所有的公共汽车候车亭以黑色和白色为主，屋顶使用不同色彩的玻璃（在Haarlemmermeer地区还设计了自行车道）。

Kerntraject Zuidtangent的14个公交车站都可被称为艺术品，那些混凝土构筑物不是毫无表情的平台。在这里，巴洛克高技派设施与必要的市民参与已经密不可分。我们花了很长时间来改进这些已经建成但未经设计的混凝土艺术品，例如，优化高架桥的断面是为了使施工更为精确，减小工程量。在所有限制条件的制约下，我们设计出了更佳的路面轮廓，使得更富有表现力的实体占据了主要地位。整个作品在候车亭彩色玻璃流动背景的映衬下特别引人注目。

在对高架桥边缘轮廓的整修中，我们使用了弯曲的钢篱笆和薄型预制混凝土板。为了让红色的Zuidtangent在连续的路面上高速行驶，就要尽量避免出现传统的梁或T型构件（至少使其不可见）。我们将边缘的支柱倾斜放置，这些由冷凝混凝土制成的大力士真正支撑住了公交专用车道。Spaarne医院高架桥的支柱高出地面，仿佛给了公交车加速行驶的动力。

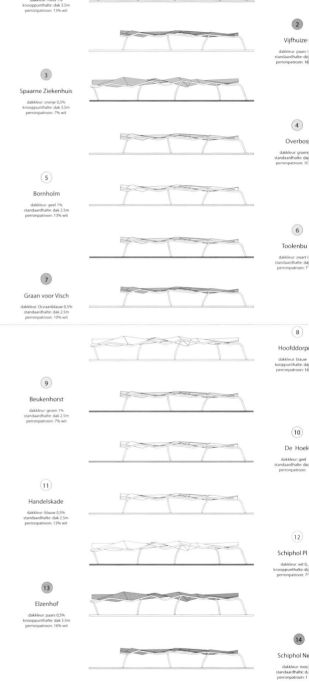

FROM THE CITIES OF NINE

A Performance on the Promenade of Blackpool

Address: Promenade, Blackpool, United Kingdom
Design: NIO Architecten
Client: ReBlackpool
Contractor: Slokker Bouwgroep bv
Structural engineer: Buro Happold
Design team: Joan Almekinders, Joost Kok, Maurice Nio, Arek Seredyn
Start design: 2006
Completion:
Costs: euro 58,000,000

来自九的城市

布莱克浦人行道上的演绎

To revitalize and animate Blackpool's new promenade takes more than a static architectural or landscape intervention. The promenade should be seen as a theatre with performers that are always active, day and night, in summer and winter. For one of the headlands (Waterloo), we have invented nine performers, nine mobile bar restaurants that populate the promenade, and nine crazy little wooden structures that ride along the coast at the speed of the strolling crowd. Sometimes these strollers gather around one centrally located restaurant, but most of time, they glide along the promenade between stops.

These nine strollers will always be there – come rain or wind – and there are also countless walkers and drifters that populate the beach and the promenade when the weather is nice, offering food and drinks to the crowds (but also keeping the coast clean and keeping an eye on things). Together with the mobile bar restaurants, they complete the colourful spectacle.

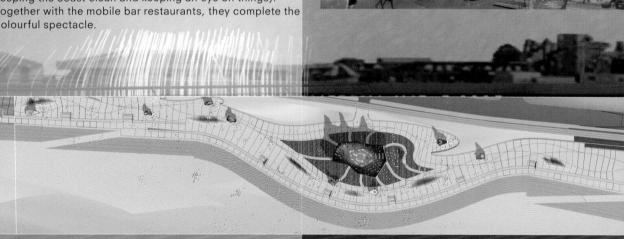

It could be regarded as a daily recurring consumption festival, but one without the sleazy flavour of a quick attraction. Here, it is not about tucking away as much fish and chips as possible, but about slowly and thoroughly enjoying what Blackpool normally does not offer. About slowly being absorbed by a strange and unknown ritual.

为了给Blackpool的海滨步道增添活力，我们不仅加入了静态建筑和景观，还将该地区视为一个永远活跃的剧场。我们在Waterloo的人行道上修建了9个移动餐吧和9个分布在海岸边的疯狂小木屋（可以和人群一起移动）。这9个小木屋有时聚集在一个餐吧周围，但大多数情况下，它们会在人行道上的站点之间滑行穿梭。

这9个推车无论刮风下雨一直停放在海滨，天气好的时候，那里会有无数的步行者和住在海滨的流浪汉，海滨步道为人们提供食品和饮料，同时保持海滩的清洁并成为海岸的焦点。再加上流动的酒吧餐馆，它们一同构成了色彩斑斓的景象。

这里已经成为日常性的消费场所，但却没有快餐式景观那种令人作呕的气味。这里不卖鱼和炸薯条，却具有Blackpool平日里少见的和缓欢乐的气氛。在这里，你会被一种奇怪且未知的仪式慢慢地感染。

GROUND CONTROL

地面控制
乌特勒支大学Uithof地球科学大楼

Faculty of Earth Sciences, Uithof, Leuvenlaan, Utrecht

Design: NIO Architecten
Client: Universiteit Utrecht
Design team: Joan Almekinders, Kristina Madsen, Maurice Nio
Start design: 2006

MIDNIGHT OIL

Cessange peripheral station, Luxemburg

午夜石油
卢森堡Cessange外环站

Address: Boulevard de Hollerich, Luxemburg
Design: NIO Architecten in comination with Polaris Architects
Client: Ministère des Transports de Luxembourg
Contractor: -
Structural engineer: Daedalus
Design team NIO architecten: Joan Almekinders, Etienne Jaunet, Maurice Nio, Giulio Piacentino, Jan Willem Terlouw
Design team Polaris Architects: Tom Bleser, Virginie Fabbro, David Gérard, Thomas Guiot, Eirik Kjølsrud, Carole Schmit, François Thiry
Start design: 2009
Completion: -
Costs: euro 150,000,000

Being independent and still being able to fit in an existing network, that is the biggest issue of today, especially for a train station. How can you be yourself, maintain your own character, while you are being tied to and in a network full of expectations and demands, and while at a few kilometres distance an ambitious main station is being designed? Maybe, by seeing the concept of "peripheral" in a new light, not as something negative, but as something positive, and that without scruples. As designers, we state this: Cessange will become a beautiful, laid-back peripheral station, without grudges. The station is unique in its kind, but at the same time, it is very laid-back with regard to all wishes and demands from the network.

That relaxed, laid-back feeling expresses itself in everything. We do not propose a monument that forces everything in its place with one gesture, but a careful unravelling and branching of streams and routes. A walk through the park is being crossed here by the course of trains, trams and busses. Thousands of commuters mingle here with students and locals every day. Offices, houses, shops and station share a collective public space. The different groups do not clash and they certainly should not avoid each other. We let them merge and stream in a new natural flow.

Not only from a logistics point of view, this station is relaxed and laid-back, also in its shape. We have carefully looked at relaxed shapes, not at canals, but at spontaneous streams of water, not at pinstripe suits, but at track suits, not at confident geometries, but at vulnerable and peripheral shapes. Hence, the seemingly unpredictable roof. Hence, the columns that are jealous of the platform. Hence, the furniture that is more Dali than Malevich. Hence, the underworld that merges everything wonderfully well, whether it is bus traffic or a simple kiss & ride moment.

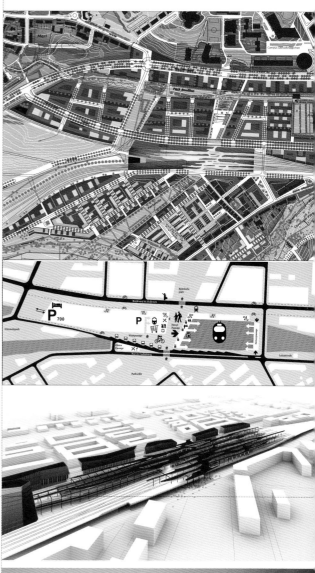

当今最为重要的问题之一是如何能在适应现有网络系统的同时保持独立，对一个火车站来说更是如此。如何能保持自己的性格，同时全面满足营运网络系统的期望和要求，在距离主火车站几公里外的地方进行设计？也许以全新的眼光看待"外围"这个概念，你会发现它积极且没有束缚。作为设计师，我们这样描述该项目：Cessange将成为一个美丽的、令人感到轻松的外围车站。该站是独一无二的，它在令人感到轻松的同时，能够全面满足营运网络系统的要求。

这种轻松、悠闲的感觉表达了建筑的一切。我们没有采用纪念碑的形式，迫使场地中的所有建筑具有相同的姿态，但我们努力避免支流和岔路的出现。穿越公园的路径在这里与火车、电车和公共汽车的路径交叉。每天都有成千上万的乘客在这里与学生流以及当地人流混杂起来。写字楼、住宅、商店和车站共享一个大的公共空间。不同的群体之间没有冲突也不彼此回避。我们让它们以一种自然的方式合并在一起。

车站的轻松和悠闲不仅体现在交通流线上，还体现在建筑的外形上。最终确定的形体不是管道型而是自由的流体，不是细条纹西装而是运动服，不是确定的几何形而是不稳定的边缘形。因此就有了看似不可预知的屋顶、令人嫉妒的平台柱列、达利风格的家具（而不是Malevich风格的）、一切事物（公交车辆、亲吻以及乘车时光）在地下都能奇妙地结合起来。

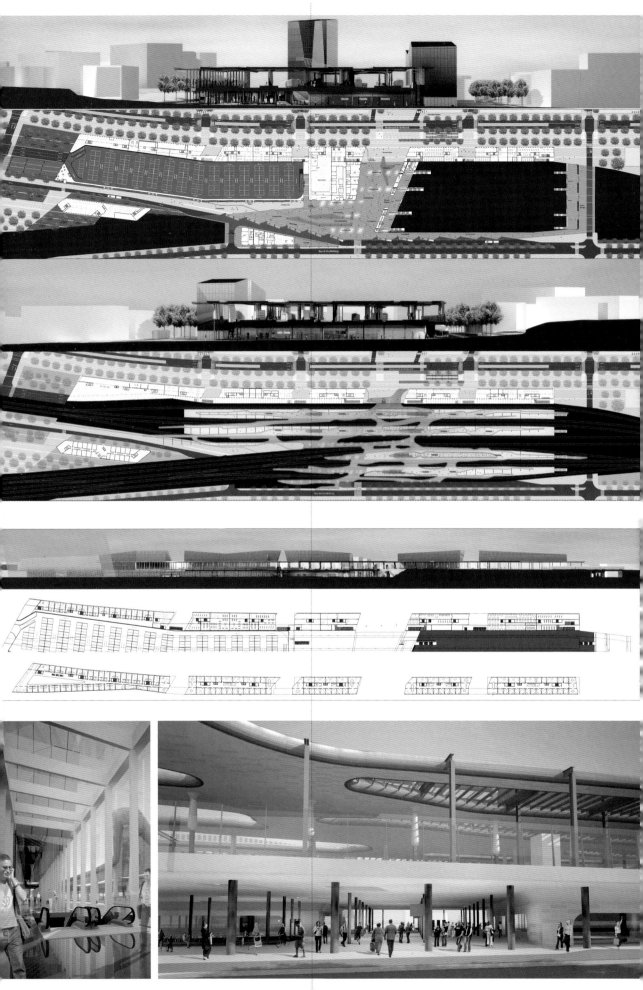

A ROSE IS A ROSE IS A ROSE

A Palace for the Queen of the Night in Rotterdam

Address: Boompjeskade Rotterdam, Holland
Design: NIO Architecten
Client: De Koninginnen van de nacht
Design team: Stefano Milani; Maurice Nio, Arek Seredyn
Start design: 2003

玫瑰呀玫瑰

鹿特丹皇后夜广场

THE THREAD OF LIVERPOOL

Bridge Restaurant in Liverpool

Address: Rocket Junction, Liverpool, United Kingdom
Design: NIO Architecten
Client: Liverpool Land Development Company
Structural engineer: Buro Happold Consulting Engineers
Design team: Joan Almekinders, Joost Kok, Stefano Milani, Maurice Nio, Alexander Paschaloudis, Wopke Schaafstal, Arek Seredyn
Start design: 2004
Costs: pound 801,506

利物浦螺纹

利物浦桥餐厅

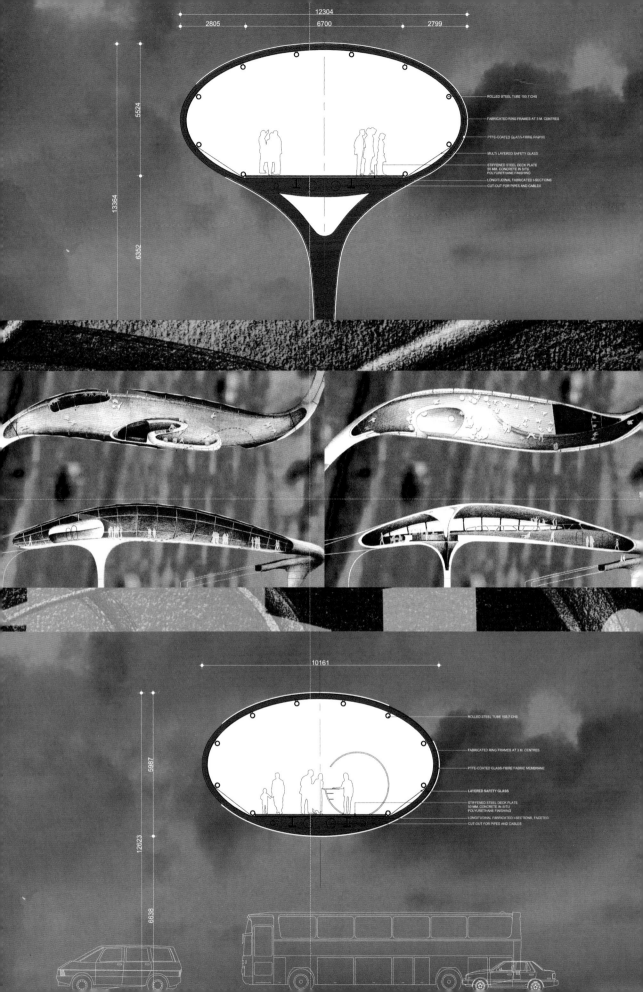

In one way or the other, in the Netherlands, we have become experts in the field of architecture in technical spaces. With "technical space", we mean all those places in our urban landscape where people normally do not hang around for long or that people sometimes cannot enter at all: dumping sites, highways, car parks, industrial estates, tunnels, viaducts, sound barriers etc. Almost all projects that we have realized and that we are working on, are related to technical spaces, and time and time again, we want to inspirit, to breathe new life into this cold, deathly, nondescript, purely functional technical space. Our device, our strategy, is to provide the soulless with a soul.

It is the same here in Liverpool. In that tangle of roads, in that desert of technical spaces, a remarkable, almost peculiar thread can be noticeable— a thin stretched red oasis. Our proposals are imbued with poetry, that is not a secret, and we are well aware of that, but it is more than lyric alone. It is a way to look at a location, a technical space, to see the possibilities there, to inspirit the place not only to make it more beautiful, but also to make it more valuable in the course of time. For some elements in our plan, the participation of inhabitants, developers and local entrepreneurs is essential to make our proposals viable.

The thread with the eight design proposals is of course, first of all, a cultural answer to the difficult question to make the entrance of Liverpool perceptible and obvious, but to us, it is more than culture alone, it should be more than purely art. In our eyes, the thread, in which the bridge restaurant plays a central role, has several faces. What was first impassable is now accessible, and what was first only reserved for cars, is now a place for social traffic as well. That is why the bridge restaurant, or bridge bar as you like, is not only a gesture to the visitors of Liverpool 2008, but just as much a destination for the inhabitants of the city.

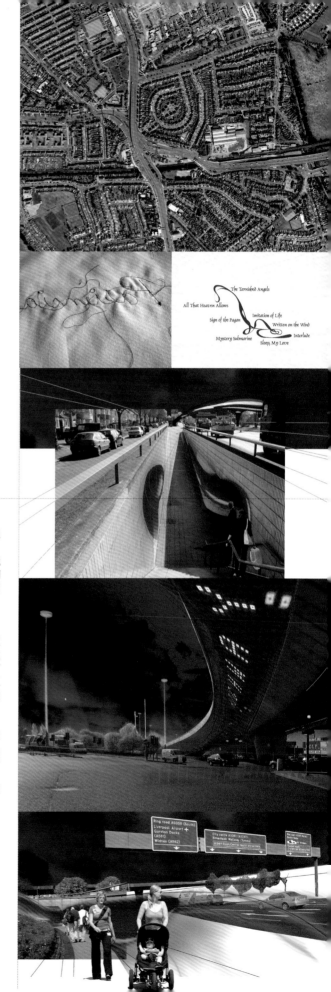

我们通过设计各种项目成为荷兰建筑技术空间领域的专家。这里的"技术空间"是指人们不常去或无法进入的城市景观：垃圾场、高速公路、停车场、工业区、隧道、高架桥、隔音屏障等地方。我们已经实现或正在设计的所有项目几乎都与技术空间有关，我们一次次地尝试给这些冰冷的、死一般的、难以归类的纯功能空间赋予生机与活力。我们要赋予它们以灵魂。

在利物浦的项目也是这样。混乱、纠结道路中有一块引人注目的、废弃的技术空间，这是一段奇特的螺纹，一段很窄的、被拉长的红色绿洲。我们受到诗歌的启发（我们都知道这不是一个秘密），但又不限于单纯的抒情，将这里看作一个技术空间，探讨基地的可能性。设计不仅要美化基地，还要提升其价值。作为一个重要的组成部分，当地居民、开发商和企业家的参与是必不可少的，他们使整个计划具有可行性。

拥有8个设计方案的螺纹首先要具有文化意义，必须使利物浦的入口清晰且明确，被世人所理解。但这还不够，设计方案不应该是纯粹的艺术。我们认为，螺纹中的桥餐厅是最重要的，该建筑有好几个立面。它疏通了原先不通行的地方，将预留车位向社会开放，它不仅是利物浦2008面对游客的门户，也是当地居民经常光顾的地方。

VOLCANO HIGH

Station Vesuvio Est, Napels

Address: Napels, Italy
Design: NIO Architecten in collaboration with Matteo Belfiore
Client: R.F.I. Rete Ferroviaria Italiana SpA
Design team NIO architecten: Joan Almekinders, Maurice Nio, Giulio Piacentino, Arjan Pit, Arek Seredyn, Fabrizio Stenti, Jan Willem Terlouw
Design team Matteo Belfiore: Matteo Belfiore, Valentina Cannava, Rocco Della Monica, Dario De Vita, Janet Hetman, Alessia Mazzei, Laura Piciocchi, Beniemino Santoro, Flavia Scognamillo, Luigi Sgueglia
Advisor Infrastructure: DHV Infra
Start design: 2009

高火山
那不勒斯Vesuvio Est车站

NAP 001 Inquadramento Territoriale

This station is not just a functional space where the travelers descend and ascend. It is especially an area that is determined by type of infrastructure systems that are connecting. Vesuvio Est station is particular because in this station cross two railway lines that differ greatly from each other: the line and the line Circumvesuviana High Speed Mount Vesuvius. The two types of lines both are literally, diametrically opposed. The first line is a hothead, but rather slow understanding; the other line is ultra mentally, but rather phlegmatic. The first line is red, while the other is blue. The station, as it reflects us, is the personification of these two different characters who meet in this place.

There is another thing that makes special design of this station is a design durable. Even where there is enough space, we do not want to waste it. The imprint was minimized and extensions will be made only above the existing building. We do not want to waste no energy. We therefore propose a building intentionally "heavy" that day remains fresh, as in churches. In the cold just a simple heating Radiant panels are sunk into the floor. This prevents the use of plants high heating energy consumption. In addition to this, we want to replace the forced ventilation systems with natural ventilation without any waste of energy.

And of course we want to avoid wasting money and time. It is absolutely impossible to achieve the two stages follow the limits set by the budget Construction of 11.5 million euros. It would be possible only if that includes structure of the final program in the first phase. By doing this, not only saving substantial time and labor costs, but also an opportunity to offer something special for the spaces of the structure that has not been determined the function could be possible. These spaces may become lush gardens where passengers and local residents can spend some time, but could also be possible to rent these spaces for local entrepreneurs to sell their products (greenhouse). It is the sort of informal market. Could think of a cross between a market and a garden, a "market garden" where you can recreate, but also be amazed by the wonder of products exposed?

火车站不只是乘客上下车的功能空间，还是各种基础设施系统连接的区域。Vesuvio Est火车站尤为如此，因为有两条差异很大的铁路线在此交汇：火车线和Circumvesuviana维苏威火山高速铁路线。这两种类型的铁路线在外观上截然不同。第一条线是一个红色的急性子，但可以被慢慢地理解；另一条线大脑迟钝且异常冷静，呈现出蓝色。该火车站是两条不同性格的线路人格化的体现。

该火车站另一个特点是建筑的可持续性。我们不想浪费建设用地（虽然空间足够大），只在现有建筑上进行改扩建。为了节能，我们特意设计了一座很"重"的建筑，在白天可以像教堂一样保持新鲜的空气，寒冷时将一个简单的加热辐射板嵌入地下，就可以避免采暖设备耗能。除此之外，我们想利用自然通风替代被动式通风系统，以达到节能的目的。

当然，我们还想节省时间和金钱。要实现以上两个设计目标是不可能将成本控制在1150万欧元的预算范围内的，除非这笔费用包含第一阶段最后的结构工程。这样做不仅可以节省大量的时间和人力成本，也可以为结构空间提供一些功能上的可能性。这些空间可能会成为供乘客和当地居民休闲的花园，也可以出租给当地企业家出售自己的产品，成为一个非正式的温室市场。是否可以将市场和花园结合起来，创造一个"市场花园"呢？也许你会惊讶于那里所出售的商品呢。

THE WAVE AND THE WHIRLWIND

ZEP Leisure Park in Middelburg

Address: Schroeweg (Mortiere), Middelburg, Province of Zeeland, the Netherlands
Design: NIO Architecten
Client: RECC/TCN Property Projects
Building contractor: Walcherse Bouw Unie
Structural engineer: SmitWesterman
Design team: Joan Almekinders, Radek Brunecky, Joost Kok, Sean Matsumoto, Maurice Nio, Arek Seredyn, Stefan Signer
Start design: 2004
Completion: 2009
Building costs: euro 27,000,000

波涛和旋风

米德尔堡的ZEP休闲公园

Photo credit: Radek Brunecky

The building for the ZEP Leisure Park is based on Zeeland's two most vital natural elements: wind and water. The image that the building evokes is that of a dance between a wave and a whirlwind. The wave is the horizontal binding theme for the events park, the whirlwind is the twenty-five-metre-high tower that can be seen from afar and that can be climbed by the daredevil.

With that the building sets the tone: not only is it embedded in Zeeland's nature, but it also refers to adventure, excitement, discovery and surprise. It is an adventurous building in which all kinds of things can be experienced, from the most intimate moment of relaxation to the wildest impressions. Everybody has something to look for here.

For this leisure park, we looked for real experiences. The trees that border the car park and that grow on the square, are real and not made of plastic. From the ZEP-tower, kids can actually float to the ground, they do not play in a virtual image. The things that are there to see, to do and to buy relate to physical, often sporty activities: skating, sailing, climbing, biking, working out, dancing and last but not least playing soccer. Of these real experiences, the building wants to be a continuation, an experience in the form of wind and water.

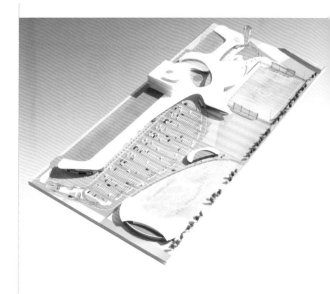

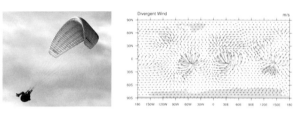

ZEP休闲公园汲取了Zeeland地区两个最重要的自然元素：风和水。形象地说是建筑唤起了波涛和旋风之间的舞蹈。波涛是主题公园水平层的主题，旋风是一座远处可见的25米高的塔，勇敢者可以攀登到塔顶。

我们这样定义建筑的基调：它不仅反映当地的自然，还要代表冒险、兴奋、发现和惊喜。在这幢充满未知的建筑物里，人们可以经历各种事情，从最亲密的放松时刻到最疯狂的事，每个人都能在这里寻找到自己想要的东西。

在这个休闲公园中，我们努力寻找真实感。停车场边界的树木和广场上的植物都是真实的，而不是塑料制品。孩子们可以从ZEP塔上飘落到地面，而不是面对一个虚拟的场景。那里的事物可以用感官去看、去做、去购买，这里充满了各种运动和活动：溜冰、划船、攀岩、骑自行车、打工、跳舞甚至踢足球。建筑想用风和水的形式延续这些实际体验。

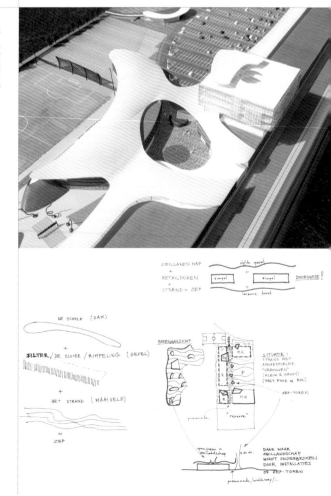

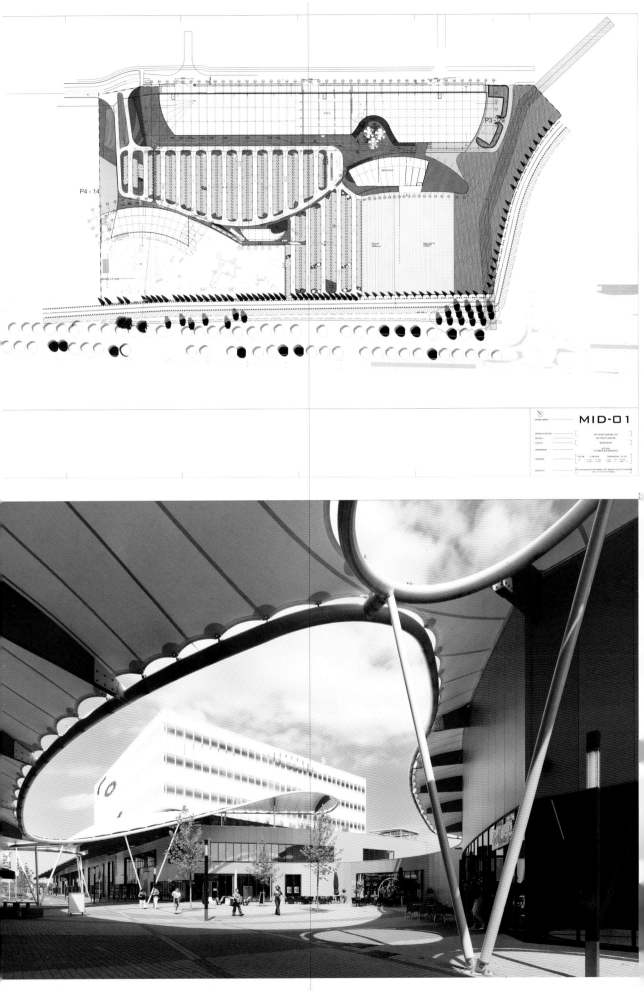

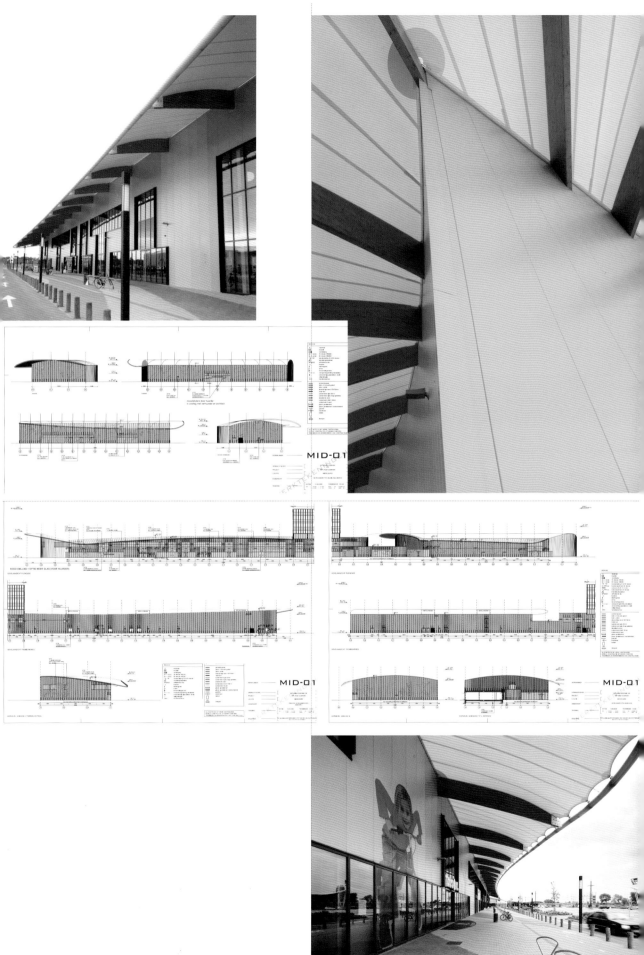

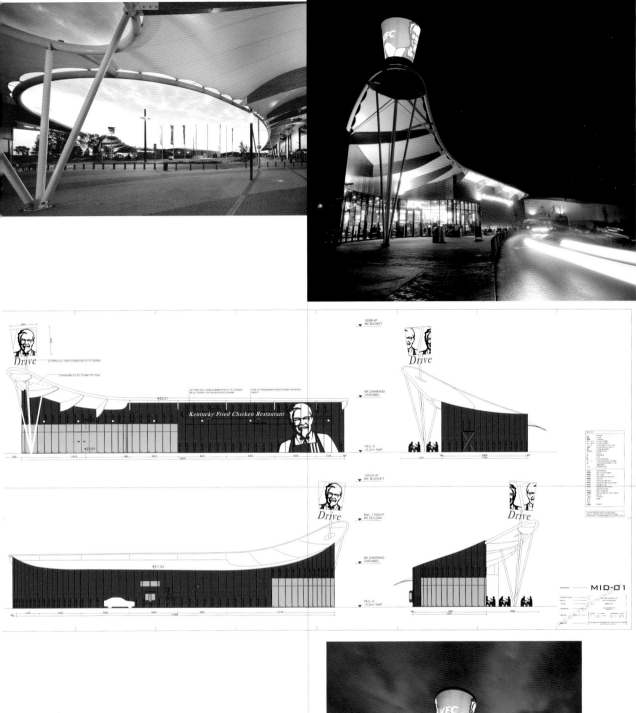
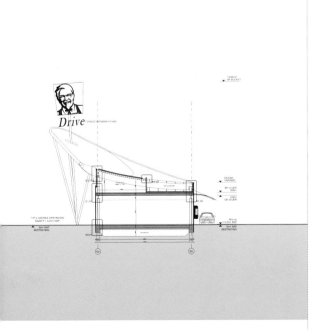

IL GIARDINO DI LIMONI

柠檬花园

Museum Extension Villa Della Regina, Turin
都灵Della Regina别墅博物馆扩建

Address: Villa della Regina, Torino, Italy
Design: NIO Architecten
Design team: Giacomo Garziano, Maurice Nio
Start design: 2010

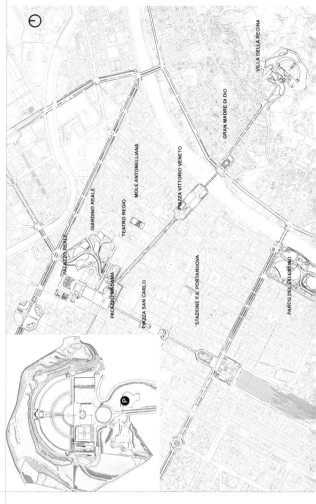
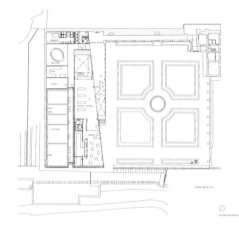
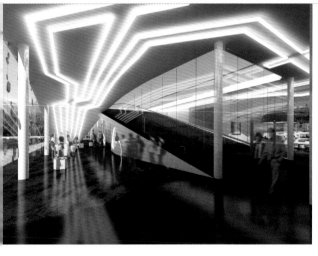

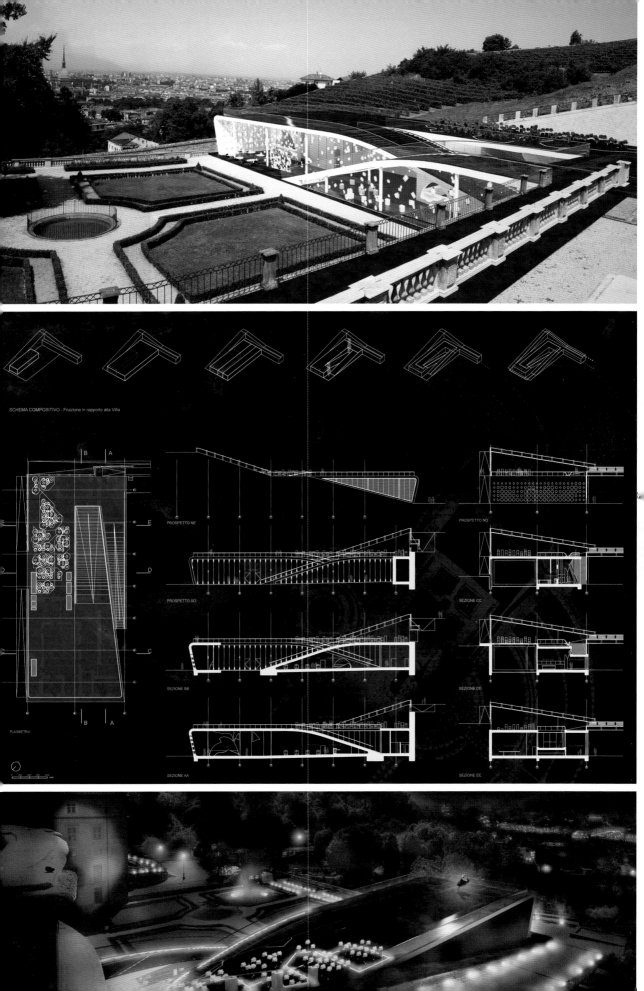

Modi Operandi.
Notes on the Work of Maurice Nio

Stefano Milani

设计方法:
对Maurice Nio作品的注释

Stefano Milani
Architect. He graduated cum laude from the I.U.A.V. of Venice. From 2001 till 2005 he had worked at Nio Architecten in Rotterdam. Since 2005 onwards he has been partner at the architectural firm Ufo Architects. He has been also carrying out a PhD research on architectural drawings theory at the Faculty of Architecture at Delft University of Technology. At the same faculty, he has also been teaching within the Territory in Transit Research Program. In 2006, he was invited to take part in the 10th Architecture Biennale of Venice. He was recently guest editor (with Marc Schoonderbeek) of the Delft School of Design Journal Footprint. He also edited the publication, Franco Purini, Drawing Architectures, Bologna 2008.

The Fire Emperor

Maurice Nio's work is characterized by many anomalous facets. Anomalous are his building and anomalous are the ways they are conceived. Anomalous is also his position within the Dutch architectural context–a context, we all know, that since the nineties have witnessed great international media recognition and that the fashionable system of architectural media progressively transformed into a global brand. Yet, the work of Maurice Nio does not quite fit in this arguably controversial scheme. Rather, his position is more the one as an underdog with a relative alterity from the-by now-overexposed phenomenon of the Dutch architecture. Moreover, he is also genuinely uninterested to engage himself in the "architectural debate" and to market his singularity as an architect. His architectural approach is too heterogeneous and unpredictable to construct a spendable aura and interest in his work. To summarize his position on architecture, we could say that if Mies' famous dictum that "you cannot invent a new architecture every Monday morning", Maurice Nio would "invent" one every day.

Active since the end of the eighties, Nio's early work with the avant-garde agency NOX, was characterized by a series of experimental editorial and architectural projects that radically investigated the possibility of the dissolution of architecture within the mainstream of the media that structure our reality and where the symbolic fiction in completely embedded within it.

After this early experiences with NOX, he further developed his particular point of view on architecture through a series of challenging project among which the AVL incinerator in Hengelo can be consider his authentic masterpiece. This work represents a sort of roman d'apprentissage for the mis au point of his "modus operandi" or, more precisely, his many "modi operandi" of the architectural project attempting to intercept marginal condition of reality, left over of the production of space and the hybridization of languages that produce or are produced by the society, by the technology, by the media. His research constantly linger upon the indeterminate space of conflicts and of the contradiction without seeking any illusory reconciliation or consolatory answer. Rather, his work aim to radicalize their dissociation, proving thus an authentic consciousness of our contemporary condition. Nio understood that is only by accepting this dimension, by performing it through a "work", by investigating the most controversial and hidden aspects of the reality that we become fully aware of it, and find the possibilities of new forms of architectural expression. The contamination of the project with external agents, the un-specificity of the media in which the projects are developed, can lead to unpredictable results and prevent the author from the tendency to repeat. In this sense the architectural project become inherently critical.

This aspect is also crucial when confronting the urban environment. Cities do not live without contaminations and conflicts-social, economical, generational, religious, and always political-that architect's main "material." It is precisely by engaging the conflicts that new conditions of the public space can emerge, and the project should constantly reveal, provoke, represent or keep them alive.

For Maurice Nio, there is no (more) substantial difference, if not dimensional, between the things and (their) image. Their semantic nature is structurally double: they are all images and all things. If architecture remains his privileged operational field, his projects exceed the common categorizations of a building to embrace the wider cultural dimension of aesthetics, of visual culture, of sociology. Conceived as representation of a myriad of other representations, Nio's projects are available for a plurality of interpretation. His buildings are actors playing a peace on a ever changing stage, or, like a David Linch' movies different "realities" coexist on the same plot. This approach have been developed on the project Amsterdam 2.0, a project for the '400 possible cities' of Amsterdam, in collaboration with the artist Paul Perry. This project can be considers his "manifesto," on the city, or a constitutional system for the dissolution of the ideal of the city. Amsterdam 2.0 in fact is not a city: it is a framework without rules for different legal systems that enable many ideal cities to coexist and overlap on the same place and on the same time.

Maurice Nio uses to "name after" his projects to references that, in various ways, played an inspiring role during the process. The common aspect of this nominalistic attitude is that he never alludes to the context or to the program addressed by the project. Rather, he only refers to fictions or fictional creatures. They can be movies titles, comic heroes, music bands, legendary monsters, names that produce a shift in perception and memory. The project refers to something we would not refer to. The name of the project "introduces" an external agent, a third party, foreign to the condition of the project. But if these heteronymous elements have little, or nothing, to do with what we would refer as the program of the project, yet their role is "necessary," as the determine a conceptual accident that breaks the one-to-one dialectic between brief and project.

Dark Matter

It is exactly this condition that Maurice pushes to the extreme till the point of reaching the absurdum. In a constant exchange, the reality is contaminated with its simulacra, the scene of the representation with the obscene, the subject with the object, the original with the copy, the beauty with the monstrous, the concrete with the abstract, the knowledge with the desire, "the intelligent with the stupid," the science with alchemy. Stylistic choices, design methodologies, theories and techniques are always adopted instrumentally and subsequently dismissed. They are used and "misused," constantly reinvented and re-elaborated to be ultimately jealously collected, when the project ends, in a sort of cabinet of curiosities. The final stage, the building, substantiated what is left from this erratic journey: the excreta of the process and measure of the architect's sovereignty.

(I worked for Maurice a number of years. These comments are a recollection of that period. As there is always oblivion in reveries, the notes above do not aim to be true or scholarly written. They are not, and they would not make much sense! I hope, they will sound al least verisimilar).

Maurice Nio的作品在很多方面都不合常理，建筑本身和设计方法都与众不同，这也成就他本人在荷兰建筑界的地位。众所周知，荷兰建筑界近九十年来取得了国际媒体的广泛认可，当前的荷兰建筑媒介系统正逐步转变为一个全球性的品牌。然而Maurice Nio的作品与这个有争议的计划并不合拍。相反，他在当前建筑界的地位更像是一个建筑主流下的失败者。此外，他本人确实对"建筑辩论"不感兴趣，也不喜欢营销自己的"与众不同之处"。在营造建筑的使用氛围和趣味性时，他的设计方法过于庞杂和不可预知。总结他在建筑界的地位，如果密斯认为："你不能在每周一清晨发明一幢新建筑"的话，Maurice Nio就可以一天"发明"一幢新建筑。

Nio自20世纪80年代末开始活跃在建筑界，他的早期作品是与前卫的NOX事务所合作的，那是一系列的实验性项目，深入调查了建筑消解的可能性。当时的媒体主流是要构建现实，象征性的虚构完全被淹没了。

通过与NOX的早期合作，Nio利用一系列具有挑战性的项目发展了自己的建筑观，位于亨厄洛的AVL焚化炉就是他的杰作。这个项目的"设计手法"是向罗马学习，更准确地说，他用许多"设计手法"捕捉现实的边际条件，当前遗留下来的空间和混杂的语言都源于社会、媒体或技术。他的研究始终围绕着相互冲突的不确定空间，不寻求任何虚幻的和解和安慰。相反，他的工作要激化它们的消解，从而证明当代真实的自觉性。Nio知道，只有接受这一事实并通过"作品"将其表达出来，调查现实中最有争议、最为隐蔽的部分，我们才能充分意识到这一点，找到新的建筑表现形式。代理商的过度参与、项目开发过程中媒体所表现出的非特异性，都会给项目带来不可预知性，导致无法复制的结果。在这个意义上，建筑本身就成为了决定性的因素。

城市环境也是至关重要的。每个城市都在社会、经济、人口、宗教、政治等多方面存在污染和冲突，这些是建筑师所面临的重要的"物质环境"。只有参与冲突才能创造新的公共空间，每个项目都应该不断揭露、促进、代表或维持这些冲突。

对于Maurice Nio来说，如果物体不是立体的，它们就与自身形象没什么重大区别。每个物体在语义上都具有双重性质：它们既是形象也是物体。如果建筑能保留自己特有的学科领域，它就可以超越普通的建筑分类，在美学、视觉文化和社会学上拥有更为广泛的文化意义。Nio的建筑旨在以点带面，为人们提供多元化的解读。他的建筑在一个不断变化的时期表现出平和的特质，像大卫·林奇的电影将不同的"现实"共存于同一情节中。这种设计方法在阿姆斯特丹2.0项目中得到了体现，该项目是建筑师与艺术家Paul Perry合作完成的。这个项目是Nio的城市"宣言"，也是为理想城市的消解所设计的法律系统。阿姆斯特丹2.0不是一个城市：它是一个为不同的法律制度设计的自由框架，在这里，许多理想城市可以在同一时空重叠共存。

Maurice Nio运用各种方法为他的项目命名，这种做法在项目实施过程中发挥了很大作用。这些名字不涉及当地文脉或项目解决方案。相反，他只会提到小说或虚构的生物，它们可以是电影的标题、漫画英雄、音乐乐队、传奇故事中的怪物等。这些事物能引起知觉和记忆的变化，却一概不涉及项目的具体内容。给项目起名是为了"引进"国外代理商和第三方，而与项目本身的条件无关。如果这些元素几乎或根本与项目计划无关，它们的"必要性"就在于打破了项目介绍和项目本身之间一对一的辩证关系。

Maurice正是将这种关系推到了极致。在这个不断变化的时代，现实总是被幻象扰乱，表达出可憎的场景，主观与客观、原创与副本、美丽与古怪、明确与抽象、知识与欲望、"智慧与愚蠢"、科学与愚昧总是混杂在一起。前期有效的解决方法是确定某种风格、运用一定的设计方法、理论和技术，但这些随后都消解了。这些风格、方法、技术和理论被人们不断地重复使用并重新解读，在项目结束时又被精心收藏起来，放在珍品陈列柜中。最终，建筑见证了这段飘忽不定的旅程所丢弃的东西：建造过程中产生的建筑垃圾以及建筑师的权利。

（我为Maurice Nio工作了数年，这些评论是对那段时光的回忆。由于难免有疏漏之处，以上文字并不力求真实，也不是学术写作，它们没有太大的意义！我希望，它们至少听上去是真实的。）

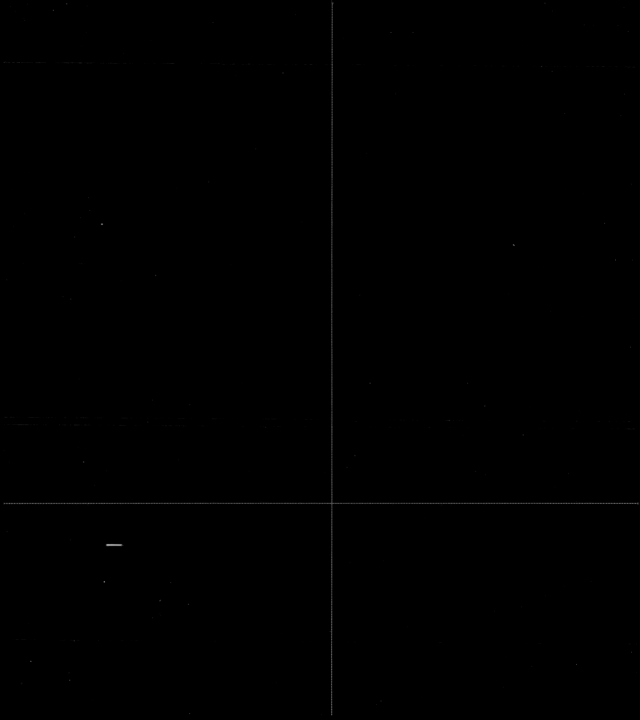

Identities
Disquieting Presences

特性
令人不安的现状

Betty Blue
Breathless
Coexistence
David and the Hulk
The Green Goblin
The Hulk
Sensing the Waves
Point Zero
Wild Orchid

BETTY BLUE

Retail Park in Roermond

Address: Sint Wirosingel, Roermond, Holland
Design: NIO Architecten
Client: Van Pol Participaties/TCN Property Projects
Building contractor: Louis Scheepers
Structural engineer: DHV Bouw en Industrie
Traffic adviser: IBZH Raadgevende Ingenieurs
Design team: Joan Almekinders, Georg Bohle, Radek Brunecky, Joost Kok, Sean Matsumoto, Maurice Nio, Arek Seredyn
Start design: 2004
Completion: 2008
Building costs: euro 20,575,000

Photo credit: Arjen Schmitz, Hans Pattist, Radek Brunecky, Hennie Retera, Marnix van Eerde, Vanpolbeheer

巴黎野玫瑰

鹿特丹大卖场

Until recently, the Dutch city centres were the stage for a shopping audience, but the last couple of years shops have been grouping together and moving more and more off-centre to develop themselves into compact shopping islands in the periphery. Despite their relatively limited size, these little shopping paradises bring about numbers of visitors that can easily compete with big amusement parks and that make the neighbouring city centres go pale. What do we offer these visitors, who until recently did their shopping in the safe surroundings of the old Dutch city centre, in which every glance into a shop window could be alternated with the well-known image of little alleys, streetlamps and hard-burnt red clinkers? In a compact setting such as this one on the 'Wirosingel' in Roermond, the audience enters a new experience, the inner world of Betty Blue, a world in which the shop and the customer communicate with each other one-to-one.

As unambiguous as this shopping machine is lying here on its doorstep, waiting for visitors, as ambiguous it is in relation to its shape and colour, it is sometimes straight and other times round, from the one side purple and from the other side blue. In the shelter of this enormous lifted and stretched drop of water, an inner square with almost exotic conditions has been shaped. It is as if a whole life of its own has been able to develop itself inside this inner space, in which façade openings, bill boards, lampposts, wastebaskets, bicycle sheds and road markings have gone through a joint and balanced growth. As if they have been able to prepare themselves in peace for years, for the arrival of hundreds of thousands of visitors and their cars, ready to host and not being interested in anything else but to treat their guests to that one, exclusive experience.

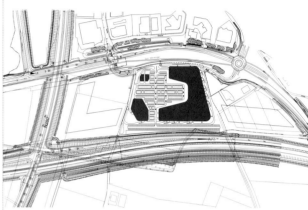

That exclusivity does not necessarily mean an extraordinary budget, is the task we set ourselves by making something with a modular and therefore efficient building system which is specific and thus unique. Where modular systems usually result in all too predictable shapes, we managed, within the regime of recurring façade elements, to put up a system of façade openings with such variation that a seemingly much bigger variety of windows, shop windows and entrance doors can be made. Even the choice for a directionless system of patterns of 8.10 by 8.10 metres did not result in a neutral building, but in a design in which it is exactly the deviations and exceptions which become visible. Is that not what everyone dreams of: a modular system that results in something unique?

两三年前，荷兰的商店还在集体迁往城外，在城市外围构成紧凑的购物区，但最近一段时间，市区里充满了购物者。尽管市内的商店规模有限，但这些小型购物天堂对游客的吸引力还是大于大型娱乐公园和郊区的购物中心。在荷兰旧城中心购物的游客能看到闻名遐迩的古老胡同、街灯和烧得火红的炉渣，什么样的商场才能把他们吸引过来呢？在Roermond的Wirosingel商场，游客将会有全新的体验，在名为巴黎野玫瑰的内部空间，游客和商场会进行一对一的交流。

这家商场就像一座购物的机器明确地坐落在台基上，等待着游客到来，与此同时，它的形体和色彩又具有模糊性：有些部分是直线，有些形体则是曲线；一面是紫色的，另一面则是蓝色的。在这个巨大的、被举起和拉伸的水滴内部，有一个引人注意的内广场。整个建筑就像拥有自己的生命一样生长出这个内部空间，这里所有的立面开口、广告牌、路灯、垃圾箱、自行车库和道路都能保持整体性并平衡发展。所有的一切看上去像是做好了站在原地的准备，等待着成千上万的游客和车辆来临，并为他们带来独一无二的感受和经历。

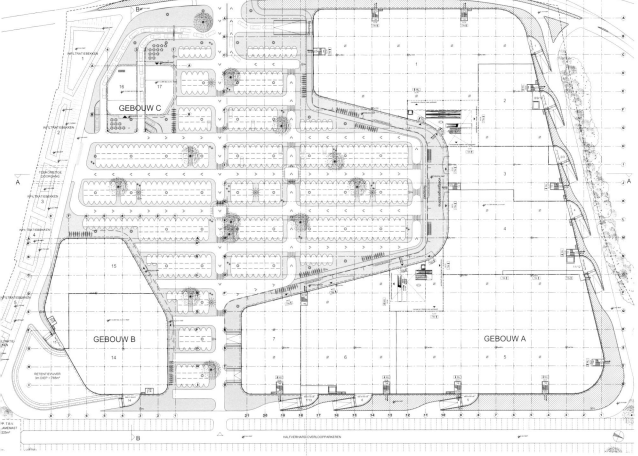

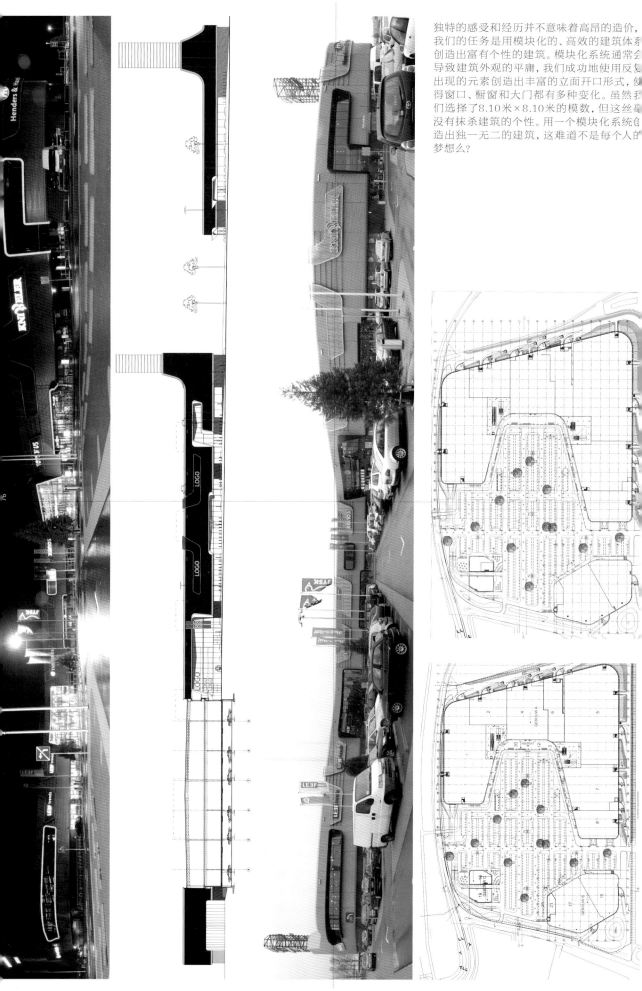

独特的感受和经历并不意味着高昂的造价，我们的任务是用模块化的、高效的建筑体系创造出富有个性的建筑。模块化系统通常会导致建筑外观的平庸，我们成功地使用反复出现的元素创造出丰富的立面开口形式，使得窗口、橱窗和大门都有多种变化。虽然我们选择了8.10米×8.10米的模数，但这丝毫没有抹杀建筑的个性。用一个模块化系统创造出独一无二的建筑，这难道不是每个人的梦想么？

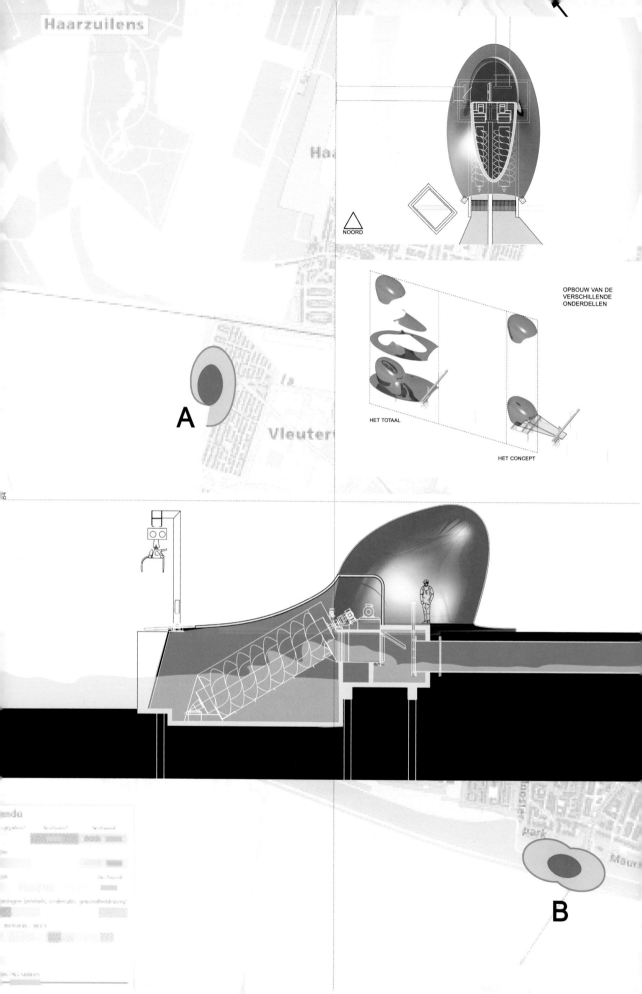

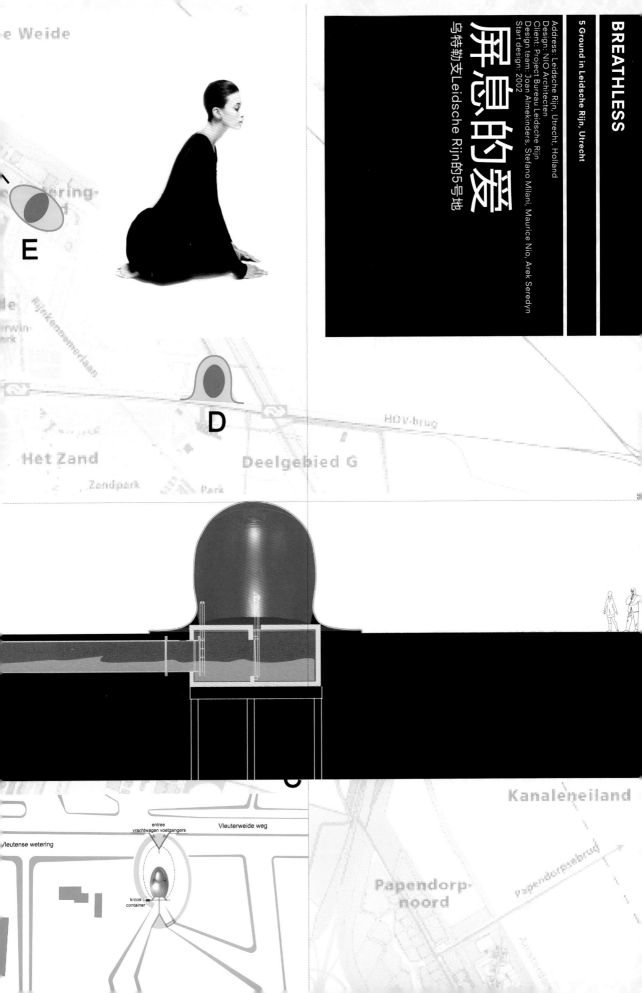

COEXISTENCE

Multifunctional Building, Den Haag

Address: Waldorpstraat, Den Haag, Holland
Design: NIO Architecten
Client: Haagwonen & Kristal
Design team: Joan Almekinders, Michal Macuda, Maurice Nio
Start design: 2008

共存体
海牙多功能大厦

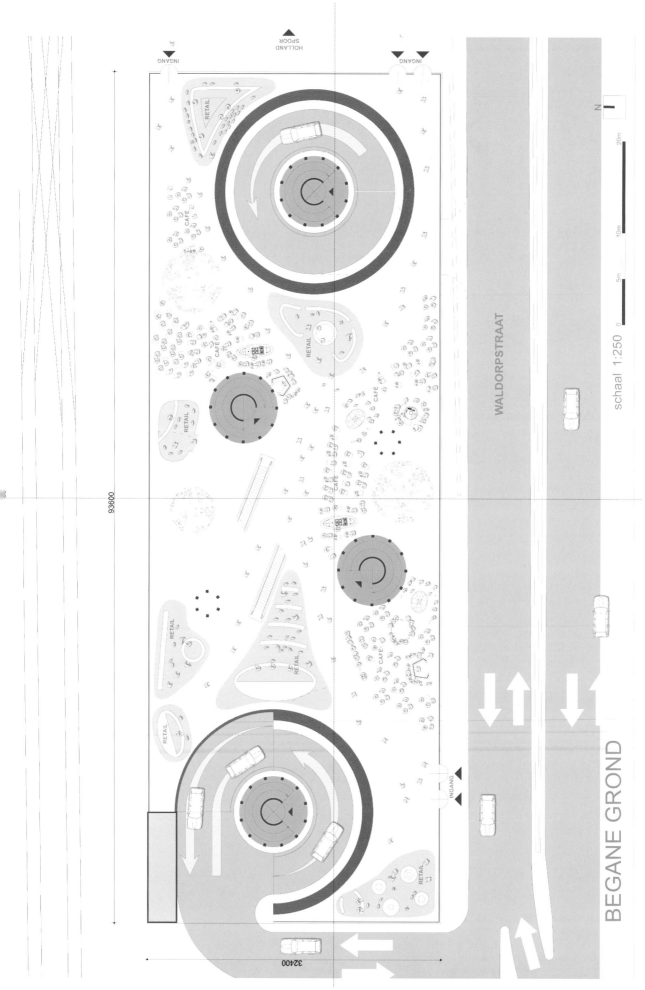

-1" VERDIEPING - PARKING
OPPERVLAKTE: 3065 m²
PARKEERPLAATSEN: 91

BEGANE GROND - LOBBY
OPPERVLAKTE: 2305 m² - LOBBY
760 m² - HELLINGBANEN

1" VERDIEPING - LOBBY
OPPERVLAKTE: 790 m² - LOBBY
640 m² - HELLINGBANEN

2" VERDIEPING - KANTOOR
OPPERVLAKTE: 1403 m² - KANTOOR
700 m² - HELLINGBANEN

TOTAL:
LOBBY:
3095 m²
HELLINGBANEN:
2608 m²
PARKEERGARAGE:
13808 m²
PARKEERLAATSEN:
400
HOTEL:
5723 m²
KANTOOR:
9060 m²
UNIVERSITEIT:
6431 m²

3" VERDIEPING - KANTOOR
OPPERVLAKTE: 1654 m² - KANTOOR
508 m² - HELLINGBANEN

4" VERDIEPING - KANTOOR
PARKEERGARAGE
OPPERVLAKTE: 975 m² - KANTOOR
1415 m² - PARKEERGARAGE
PARKEERPLAATSEN: 38

5" VERDIEPING - KANTOOR
PARKEERGARAGE
OPPERVLAKTE: 1050 m² - KANTOOR
1415 m² - PARKEERGARAGE
PARKEERPLAATSEN: 38

6" VERDIEPING - KANTOOR
PARKEERGARAGE
OPPERVLAKTE: 1023 m² - KANTOOR
1415 m² - PARKEERGARAGE
PARKEERPLAATSEN: 38

7" VERDIEPING - KANTOOR
PARKEERGARAGE
OPPERVLAKTED: 985 m² - KANTOOR
2166 m² - PARKEERGARAGE
PARKEERPLAATSEN: 65

8" VERDIEPING - KANTOOR
PARKEERGARAGE
OPPERVLAKTE: 985 m² - KANTOOR
2166 m² - PARKEERGARAGE
PARKEERPLAATSEN: 65

9" VERDIEPING - KANTOOR
PARKEERGARAGE
OPPERVLAKTE: 985 m² - KANTOOR
2166 m² - PARKEERGARAGE
PARKEERPLAATSEN: 65

10" VERDIEPING - HOTEL
UNIVERSITEIT
OPPERVLAKTE: 1163 m² - HOTEL
1393 m² - UNIVERSITEIT

11" VERDIEPING - HOTEL
UNIVERSITEIT
OPPERVLAKTED: 1055 m² - HOTEL
1280 m² - UNIVERSITEIT

12" VERDIEPING - HOTEL
UNIVERSITEIT
OPPERVLAKTED: 1321 m² - HOTEL
1029 m² - UNIVERSITEIT

13" VERDIEPING - HOTEL
UNIVERSITEIT
OPPERVLAKTED: 1117 m² - HOTEL
1301 m² - UNIVERSITEIT

14" VERDIEPING - HOTEL
UNIVERSITEIT
OPPERVLAKTED: 1067 m² - HOTEL
1428 m² - UNIVERSITEIT

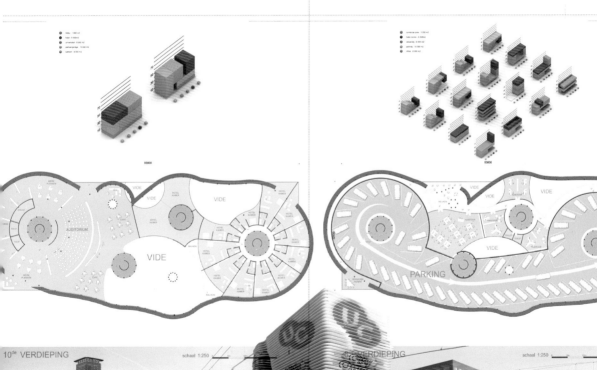

10ᵈᵉ VERDIEPING   schaal 1:250   VERDIEPING   schaal 1:250

DAVID AND THE HULK

Extension office building Twence Inc. Waste Processing in Hengelo

Address: Boldershoekweg 51, Hengelo, Holland
Design: NIO Architecten
Client: Twence bv Afvalverwerking
Contractor: Bouwbedrijf Punte/Systabo Systeembouw
Structural engineer: BDG Architekten Ingenieurs
Design team: Remco Arnold, Mark Bitter, Maurice Nio, Arek Seredyn, Jaakko van 't Spijker
Start design: 2000
Completion: 2003
Costs: euro 2,150,000

Photo credit: Hans Pattist

大卫和绿巨人

亨厄洛Twence公司垃圾处理场办公楼扩建

Ever since the delivery of the waste processing installation in 1997, the building has not changed. But after the organizations aviTwente and RegioTwente (environmental sector) merged in 2000, there was an immediate need to double the office floor area. This extension also offered the possibility to find a solution for the wish to connect the office building with the main building and the installation, and thus to even out the psychological barrier between white collars and blue shirts.

In order to make a clear distinction between the big waste processing installation and the relatively small extension a different design and use of material is chosen. Compared to the expressive main building, the design of the new office building is quite flat and two-dimensional. The point of departure even being a line: the corridors on which office spaces border alternately left and right. This line starts on top of the existing office building, descends to first floor level then to rise again and to connect to the 7.2 meter level of the waste processing installation. This is the level on which all machines are installed and where the central control room is.

Also for the finish of the façade, a flat look is chosen. As opposed to the metallic green surfaced façade plating, pearl white smooth sandwich panels is chosen. That is why maybe the extension has become more wayward than the main building, a slender snakelike shape that dares to besiege the big waste processing installation, a small white David besieging the strapping green Hulk.

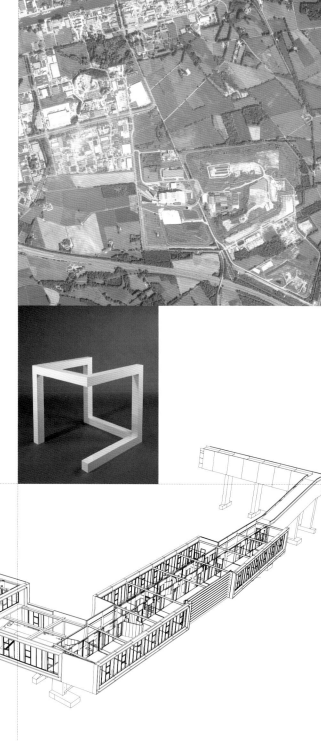

自从1997年垃圾处理分厂交付使用后，这座建筑就没有改变过。但在aviTwente组织和RegioTwente（环境部门）于2000年合并后，这里的办公面积就迫切需要变为原来的两倍。改扩建工程还使得主楼和办公楼有可能连接起来，以打破白领阶层和蓝领阶层之间的心理障碍。

为了明确区分大型垃圾处理设备和相对较小的扩建部分，我们采用了不同的设计策略并选用了不同的材料。相对于具有表现力的主体建筑，新办公楼的设计相当平缓甚至趋于平面化。我们的出发点是使走廊上办公空间的边界左右交替。这条线开始于现有办公楼的顶部，下降到一楼后再次上升，最终连接到垃圾处理设备7.2米的高度。所有机器都安装在这一高度，中央控制室也位于此。

最后完成的立面同样采用了扁平的外观。与绿色电镀金属表面相对的是珍珠白的光滑夹芯板。这样一来，扩建部分看上去比主体建筑更有个性，修长的蛇形物体竟然缠绕在巨大的垃圾处理设备上面，就像一个白色大卫缠绕住了绿巨人。

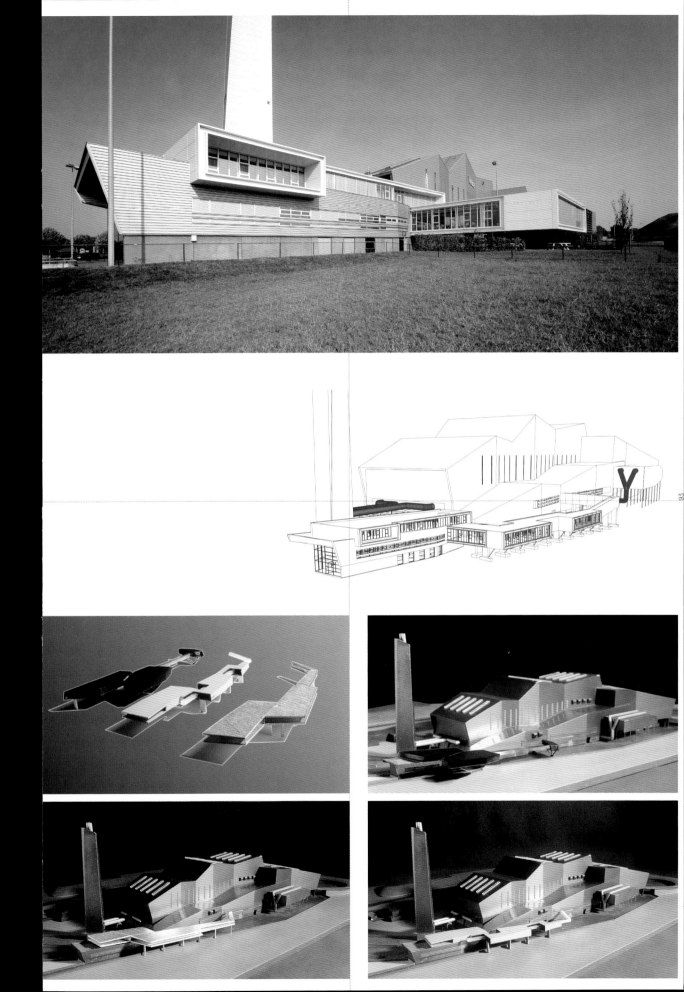

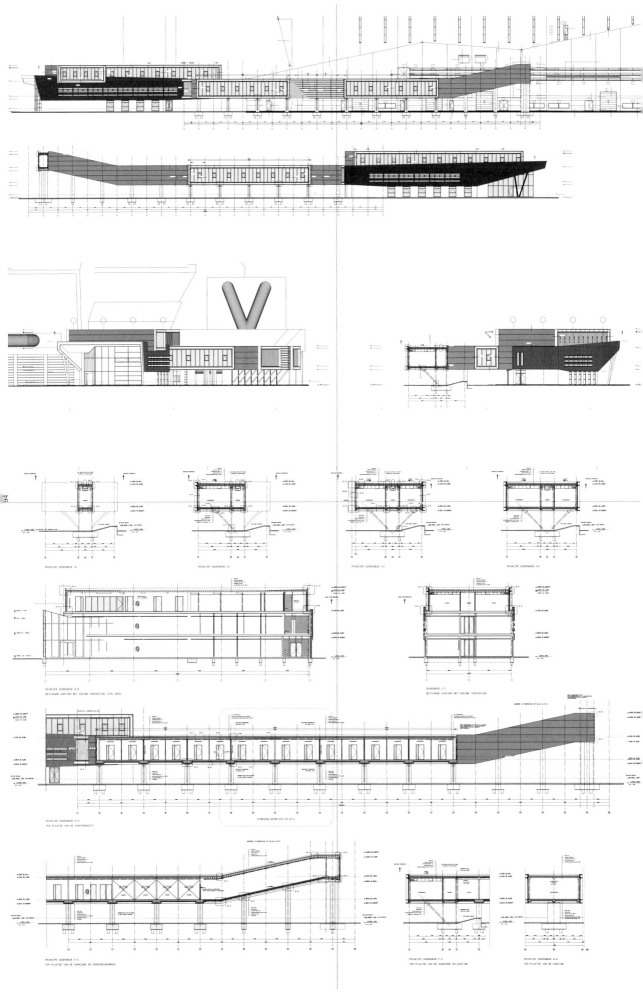

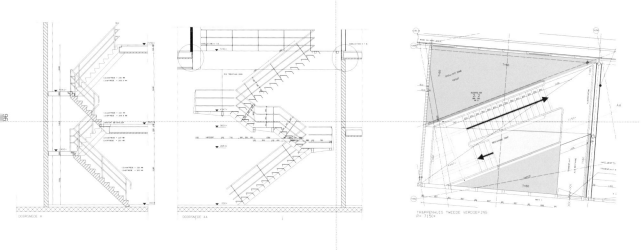

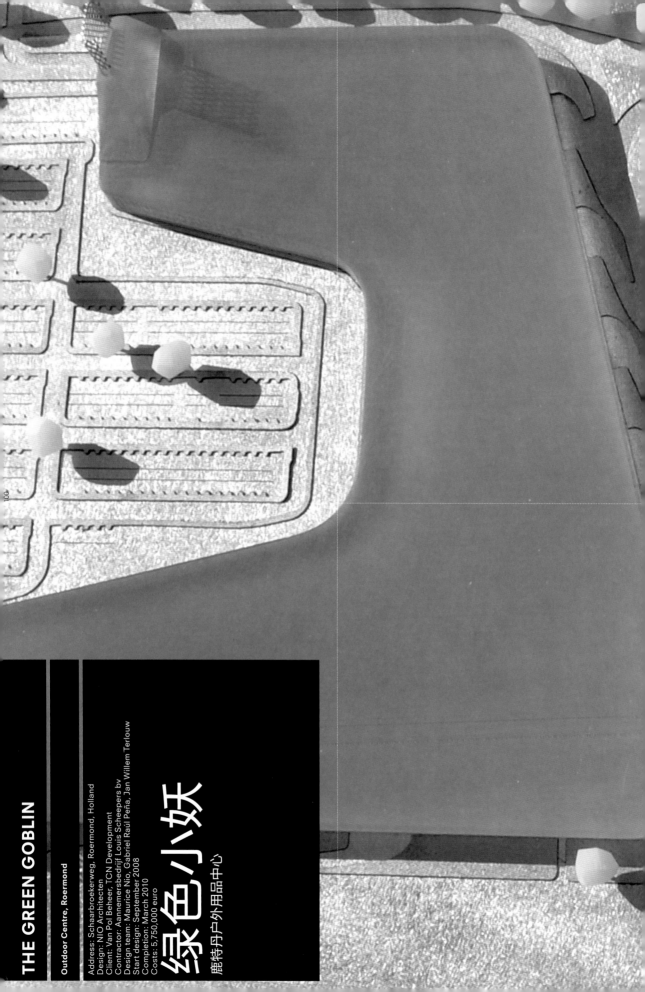

THE GREEN GOBLIN

Outdoor Centre, Roermond

Address: Schaarbroekerweg, Roermond, Holland
Design: NIO Architecten
Client: Van Pol Beheer, TCN Development
Contractor: Aannemersbedrijf Louis Scheepers bv
Design team: Maurice Nio, Gabriel Raúl Peña, Jan Willem Terlouw
Start design: September 2008
Completion: March 2010
Costs: 5,750,000 euro

绿色小妖

鹿特丹户外用品中心

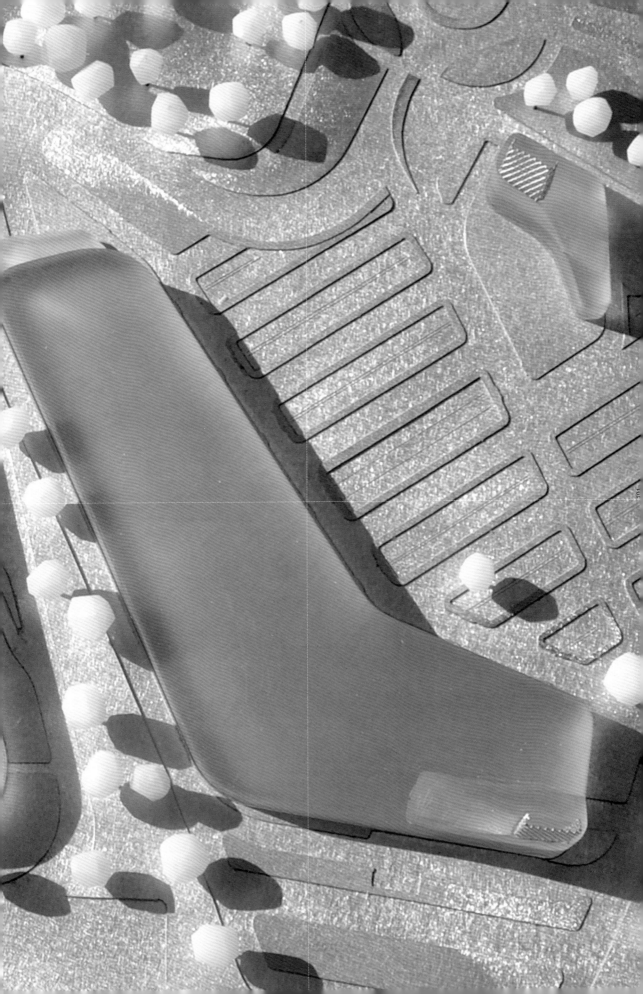

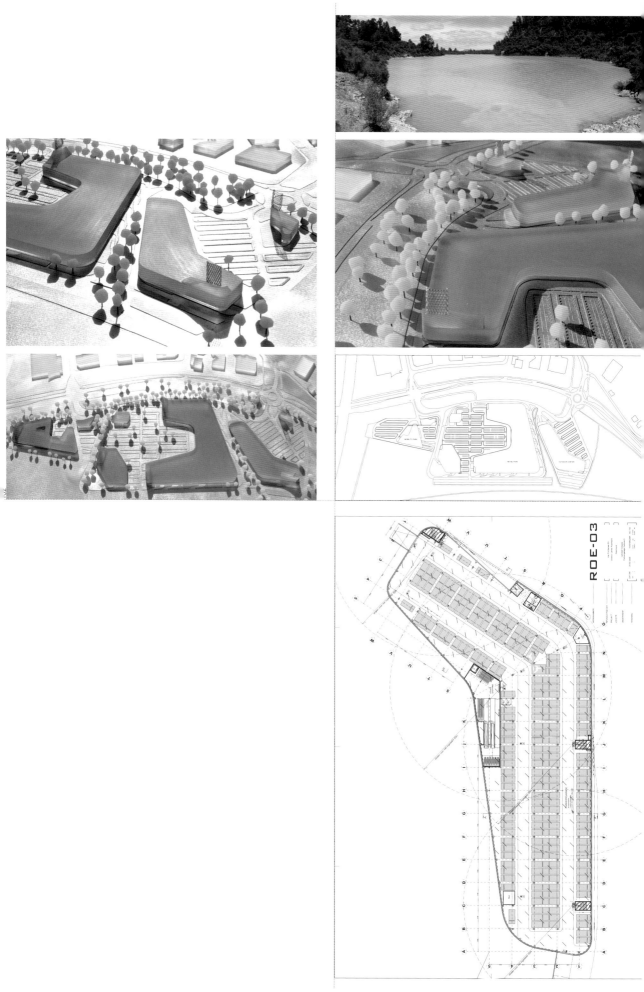

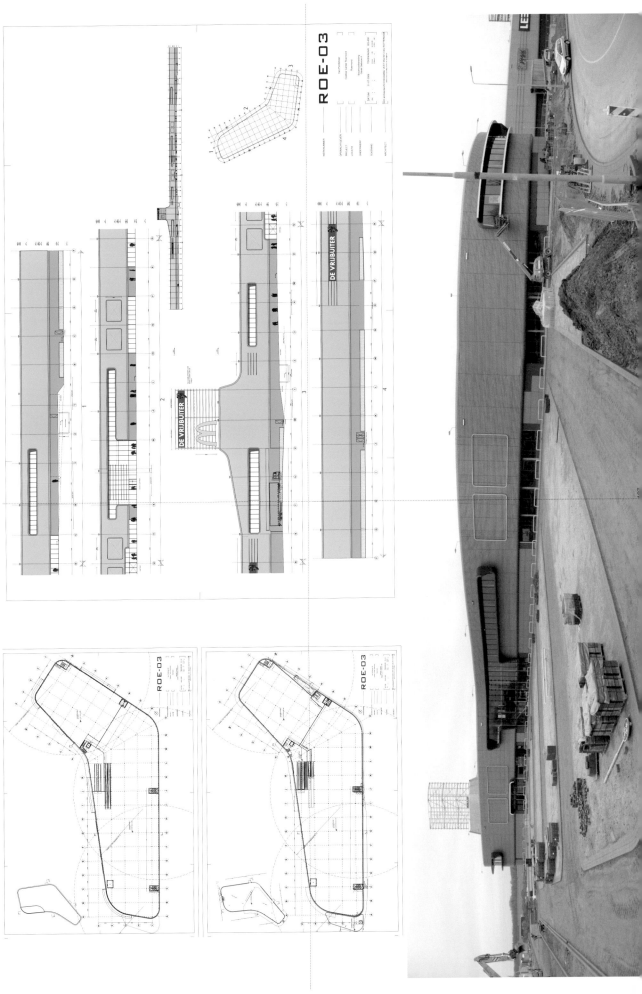

THE HULK

Waste incineration Plant in Hengelo

Address: Boldershoekweg 51, Hengelo, Holland
Design: BDG Architekten Ingenieurs
Client: aviTwente bv
Contractor: BAM/Bredero
Structural engineer: BDG Architekten Ingenieurs
Design team: Maurice Nio, Ernst van Rijn
Start design: 1993
Completion: 1997
Costs: euro 270,000,000 (installations included)

绿巨人
亨厄洛垃圾处理场

Photo credit: Hans Pattist

In conceptual terms, the waste incineration plant in Twente is the filter between the rough-hewn landscape surrounding the city of Hengelo and the advanced installation system within. The building works as a kind of interface between the external and internal landscape, between the coarse and the subtle, between the past and the future of rubbish. The building has assumed the features of the rubbish tips outside and the characteristics of the invisible incineration and decontamination process.

This is because the two independently functioning incineration lines are essentially a digestive process. Both lines have to incinerate 230,000 tons of rubbish each year and to decontaminate flue gases. This digestive process most closely resembles that of a human being or an animal. And that is why this process inspired us to make an insect, a metallic green dung beetle with mechanical features, a beetle that feeds on all that rubbish and excretes clean flue gas.

We allowed this idea, this theme, to reverberate throughout the entire structure so that it resounded in the main building and in the five secondary buildings on the grounds which consist of the water and air-cooled condensers, the slag treatment building, the weigh house, the office building and the gas reduction station. But beyond that, it is also echoed in the very details and finish of the site, and in the many pieces of furniture that populate the building like small insects.

在概念上，Twente垃圾处理厂将粗犷的Hengelo市郊景观过渡到先进的垃圾处理设备。该建筑是外部和内部景观、粗糙和微妙、垃圾的过去与未来之间的接口。这座建筑体现了垃圾处理过程中外部倾倒、不可见焚烧以及净化过程的特点。

处理厂内部的两条独立的焚烧线进行着垃圾的焚化过程。一条焚烧线每年都要处理至少23万吨垃圾，并净化产生的烟气。这与人或动物的消化过程很相似。我们由此获得灵感，将建筑设计成一只昆虫，一只拥有绿色金属外壳和机械特性的甲壳虫，一只以垃圾为食并净化烟气的甲壳虫。

我们将这个主题贯穿在整体布局中，基地上分布着主体建筑和五幢附属建筑（水和空气冷凝器、废渣处理楼、垃圾称重处、办公楼和气体减排站）。除此之外，这一主题也体现在细节和对基地的处理上，建筑中的很多家具也设计成昆虫的形状。

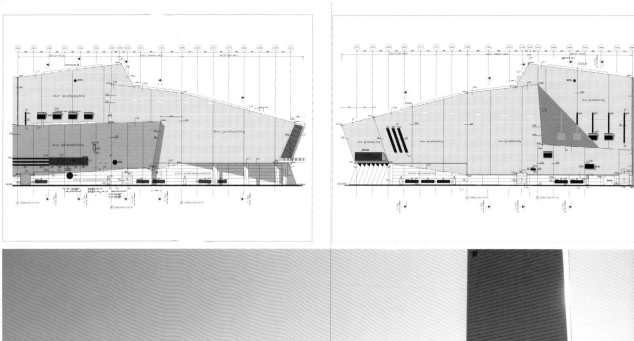

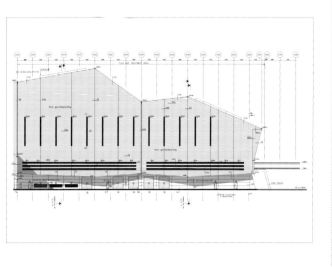

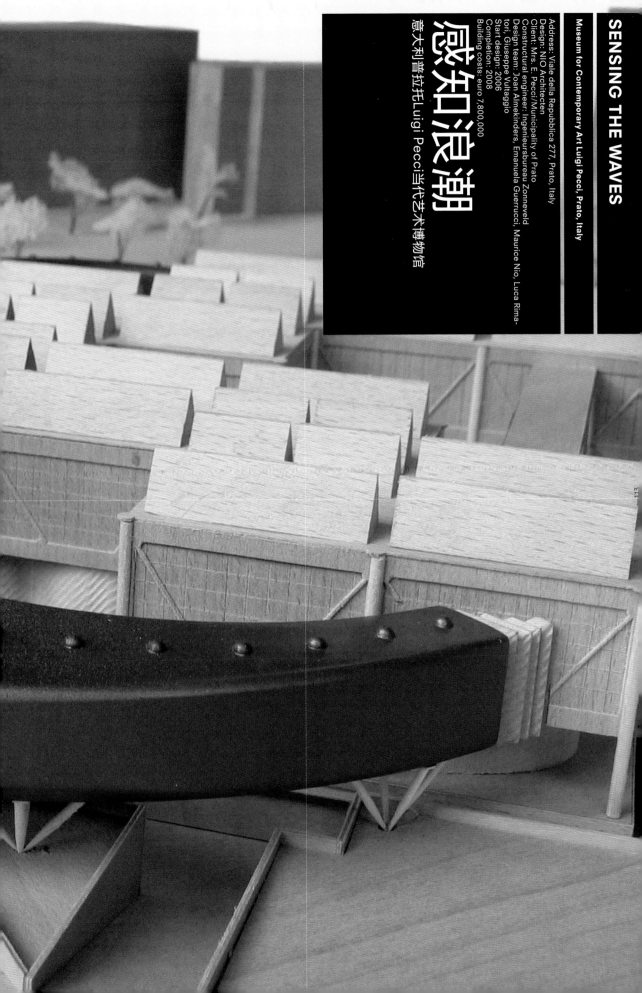

SENSING THE WAVES

Museum for Contemporary Art Luigi Pecci, Prato, Italy

感知浪潮
意大利普拉托Luigi Pecci当代艺术博物馆

Address: Viale della Repubblica 277, Prato, Italy
Design: NIO Architecten
Client: Mrs. E. Pecci/Municipality of Prato
Constructural engineer: Ingenieursbureau Zonneveld
Design team: Joan Almekinders, Emanuela Guerrucci, Maurice Nio, Luca Rimatori, Giuseppe Vultaggio
Start design: 2006
Completion: 2008
Building costs: euro 7,800,000

The "Centro per l'arte contemporanea Luigi Pecci" was opened in 1988 and donated to the city of Prato by Enrico Pecci, in memory of his son who died at an early age. The museum is situated on the periphery of Prato, near the exit of the A11 highway, a strategic spot where, from the first floor, you can see the skyline of Florence, the city where tourism and ancient culture reign. On this spot, however, two opposites dominate: (textile) industry and modern art. The art centre is one of the few museums in Italy that is devoted to modern art and furthermore, that possesses a superb collection which, for lack of exhibition space, is stored in various depots. To be able to display the invisible works of art, it was decided to double the exhibition space and to solve two important problems with the new construction.

One problem is that it is not possible to make a tour through the museum, there is a route, but that is linear (when you arrive at the end, you have to take the same way back). The other problem is that no one can find the entrance. It looks just like the imperial palace in Tokyo, super visible, but inaccessible. The first problem was solved by creating a circular plan on the first floor, where all current exhibition rooms are, in such a way that several tours can be made. The second problem was solved by situating all public services on the ground floor and by explicitly orienting the main entrance towards the street.

As opposed to the rather rigid, mechanical character of the existing museum building, partly inspired on the industrial textile markets in Prato, the new part looks fluid and ecstatic. It embraces the existing building and touches it only there where needed for the circular plan. Because the cross section of the exhibition floor constantly changes, within the interior different spaces with different atmospheres come into being, and thus different exhibition possibilities. The tower is a story on its own. It is a crossing between a horn and a feeler: on the one side, it is a weapon that is proudly presented to the visitors and passers-by and on the other side, the tower senses conditions that are immeasurable for radars and people. It gauges the cultural mood, in search of new movements.

Enrico Pecci为了纪念自己早年夭折的儿子,于1998年将"Luigi Pecci当代艺术中心"捐赠给普拉托市。该博物馆坐落在普拉托郊外,靠近A11公路,从那里可以看到以旅游业和文化产业而闻名的佛罗伦萨。而在普拉托市,主导产业却是纺织业和现代艺术。该艺术中心是意大利为数不多的展示现代艺术的博物馆,此外由于展览空间不足,很多精美的展品都存放在仓库中。为了更好地展示艺术品,当地政府决定加建博物馆,达到比原来大一倍的展览空间,并解决以下两个重要问题。

首先,博物馆只有一条线性的参观流线,游客参观结束后必须原路返回;另一个问题是,没有人能找到建筑入口,整个建筑看起来就像东京皇宫,可望而不可即。我们在一层设计了一个圆形大厅,这样,所有的现有展厅就都有多种进入方式;针对第二个问题,我们将所有的公共服务设施放置在一层,并明确了面向街道的主要入口。

面对僵化呆板的现有博物馆建筑,我们受到普拉托纺织工业市场的启发,赋予它流动的、令人惊喜的外观。新建部分包围了现有的建筑,与圆形平面的衔接堪称点睛之笔。由于建筑的剖面不断地变化,我们在建筑内部营造出具有不同气氛的室内空间,为举办不同的展览创造可能性。博物馆的塔寓意独特,它位于两种不同空间感觉的交叉点,一边令人感到武器般的高傲,另一边则让人感到无法掌控的神秘感,这座塔把握住了某种文化情结,寻求着新的变化。

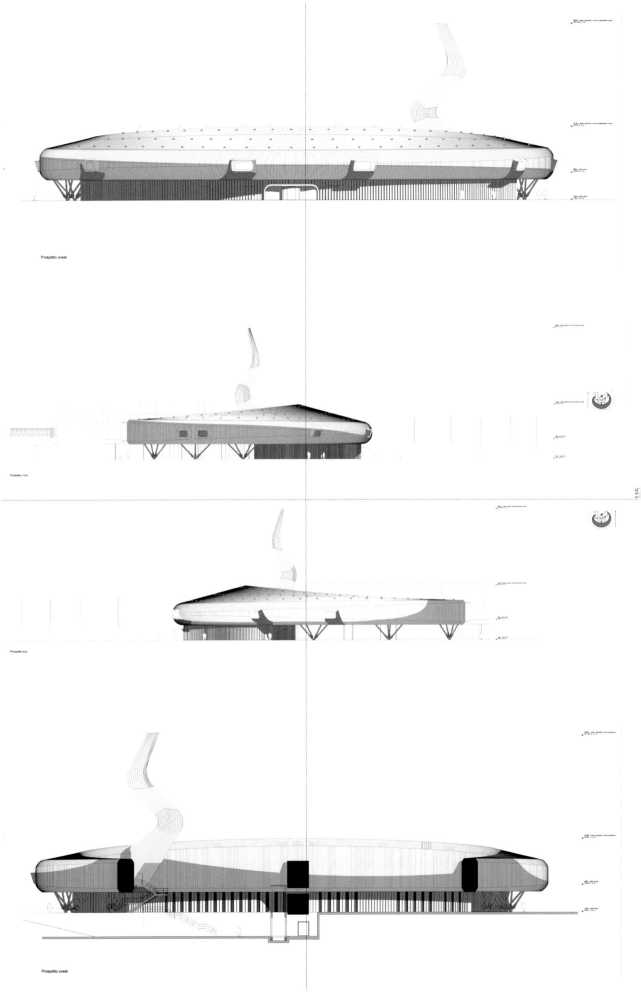

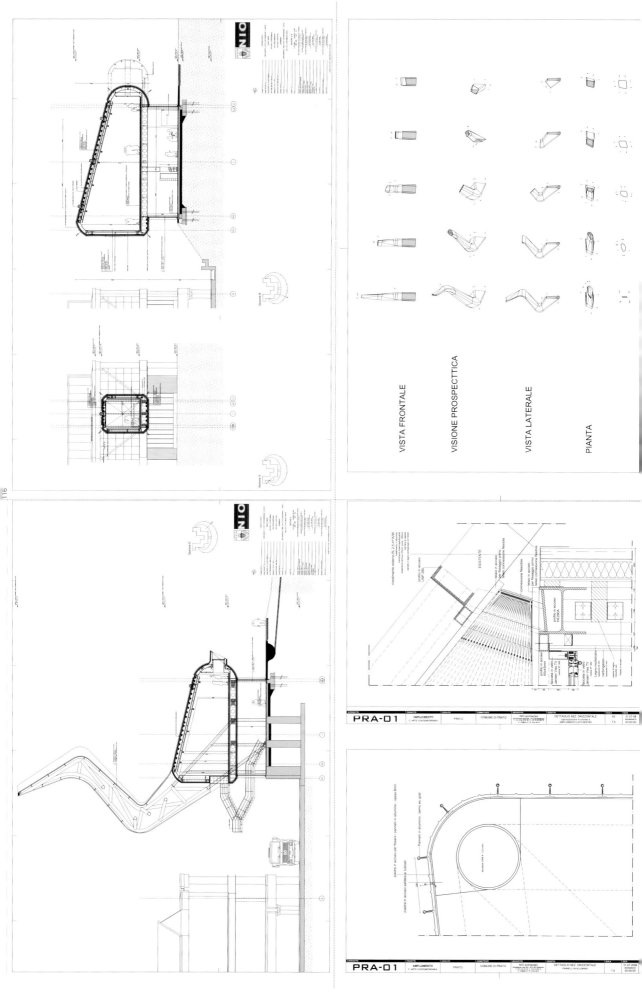

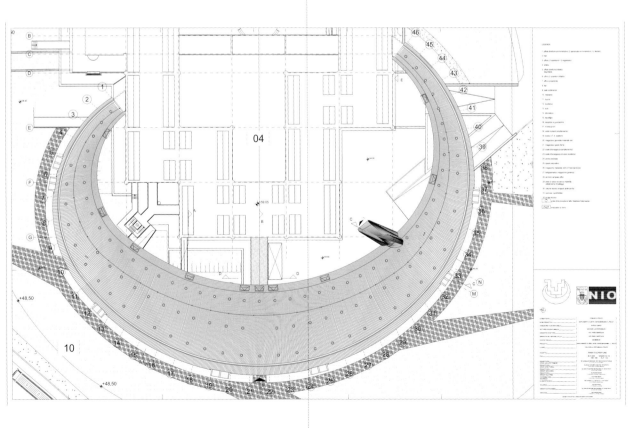

POINT ZERO

Houseboat in Amsterdam

Address: Holendrechterweg, Amsterdam, Holland
Design: NIO Architecten
Client: Maarten Aris, Amsterdam
Design team: Joan Almekinders, Maurice Nio, Alexander Paschaloudis
Start design: April 2005

零点 阿姆斯特丹船屋

This houseboat is located in the blind spot of Amsterdam. Where the idyllic little river "de Holendrechter" crosses the broad A2 motorway, the houseboat is moored. A seemingly impossible spot that could have come out of a book by Ballard: on the one side, the lovely landscape of water, little boats and reeds and directly on the other side, the deafening scene of the motorway, trucks and, at night, the accompanying pale, yellow light. It is a spot where light is never turned off, just like sound. This is also an impossible spot legally, because the boat is actually not supposed to lie here, it is being tolerated, until its occupant passes away.

This houseboat is located in the blind spot of Dutch house building. Architects always nurse hopes that the people who are looking for different types of housing – in empty offices, industrial buildings that have become obsolete (lofts), old farms, holiday homes, caravans and houseboats – will gain the upper hand a little bit more and that the typical housing format which is frozen in the brains of the estate agents, will start to melt. But highly likely this houseboat is once again an example that the gap between specific individual living desires and the serial development of house building with its accompanying rules, is too big to fill.

This houseboat is located in the blind spot of architecture. All architectonic references have been soaked off, peeled off, reduced, until something remains that does not know so well what it is: possibly something amorphous, maybe still something defined. Not that this makes it insecure or unhappy, on the contrary, it cultivates its ambiguity deep into its interior. No form or function wants to be sure of itself and has become unrecognizable in relation to the inevitable furniture, just like the houseboat itself in relation to architecture.

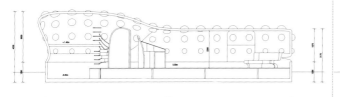

这座船屋位于阿姆斯特丹的一个盲点,它停泊在田园诗一般的de Holendrechter小河与宽阔的A2高速公路的交叉点上。在巴拉德的书中,这是一个看似不适于生活的地方:河的一边是渔船和芦苇构成的美景,另一边则是嘈杂的高速公路、源源不断的卡车和晚上淡黄色的灯光。这里的灯光从未熄灭,噪声也不绝于耳。而且这座建筑本身就是不合法的,因为这里本不该停船,一旦船的主人去世,船屋就会被拖走。

这座船屋位于荷兰住宅建设的盲点。建筑师总是希望客户能超越传统的住房观念,寻找不同类型的住房,如空置的办公室、废弃的工业建筑(阁楼)、旧农场、度假屋、大篷车和船屋等。但是,这座建筑中个人生活需求和住房发展(伴随着相关法规的发展)间的差距太大,难以填补。

这座船屋在建筑上也处于盲点的位置。所有建筑上涉及的东西都被浸泡过、提炼过、简化过,直到最后不知剩下的是什么:也许是一些无定形的东西,也许是一些确定的东西。这些东西并没有破坏建筑的安全性,引发人们的不悦;相反,它们赋予建筑以内在的模糊性。任何与必备家具有关的形式或功能都是不确定而难以识别的,就像船屋本身和建筑的关系一样。

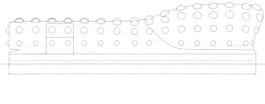

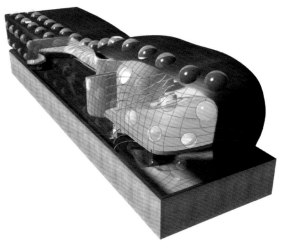

WILD ORCHID

Villa in Waalre

Address: Waalre, Holland
Design: NIO Architecten, Rotterdam
Client: Private person
Design team: Joost Kok, Maurice Nio, Arek Seredyn
Start design: 2006

野兰花
瓦尔勒别墅

We start downstairs, in the basement. In this rather quirky space with unusual alcoves, twelve extraordinary and very expensive cars are being displayed, and on the walls is a collection of works of art. This is the space for the collection.

A secret little staircase connects the basement with the house above it, on the ground floor, a continuous space shaped like an orchid. There actually aren't any rooms. Private rooms could be created, however, in the form of objects in the space – a bedroom object, a kitchen object, a bathroom object – but essentially the house is one space that is shared by everybody, just like the villas of Palladio in which everyone – parents, kids, guests, servants – uses the same space. This is the space of the family.

A perky spiral stair opens up the roof. A wild rib structure is decoration and construction at the same time. This is a space where the family wants to give parties, on the roof, surrounded by trees. This is the space for the guests, and for heaven.

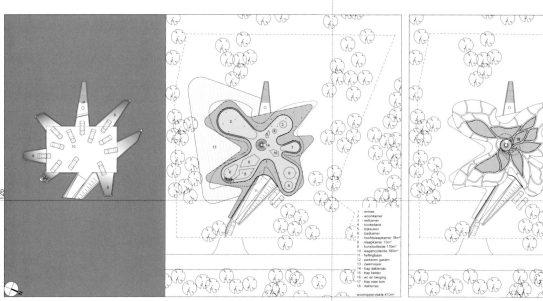

我们的建筑之旅从地下室开始。在这个古怪的空间中，有一个不同寻常的壁龛、12辆昂贵的车，以及挂满墙壁的艺术作品。这里是别墅的储藏空间。

一个秘密的小楼梯连接着地下室和一层空间。一层的连续空间形状犹如一朵兰花，这里没有房间的划分。可以用家具划分出个人空间，例如，卧室、厨房、浴室等。但在本质上，这座房子是一个被大家共同分享的大空间，正如帕拉迪奥别墅一样，每个人（父母、孩子、客人、仆人）都享有同一个空间，这就是家庭的空间。

沿着一段富有生气的螺旋楼梯拾级而上可到达屋顶。这个天然的骨架既是装饰也起到结构的作用。屋顶上的空间被树木环绕，这里是家庭聚会的地方，是为客人和天空准备的。

Building Fictions
Hans Ibelings
(Translation: Julia van den Hout)

虚构的建筑

Hans Ibelings
Hans Ibelings (Rotterdam 1963) is editor of A10 new European architecture, a magazine which he founded in 2004 together with Arjan Groot. He is the author of several books, inlcuding Supermodernism: Architecture in the Age of Globalization and European architecture since 1890.

Bus Stop- "Flower Power"

The most intriguing works by Maurice Nio are interventions in places where normally little architecture is found. Bus stops, viaducts, a shopping mall and a waste disposal plant can often exist without much architecture, and Nio's work shows that such environments also can live with a great architectural richness.

Nio's aesthetic intensification of ordinary places has an alienating effect, especially because you do not expect it. As in a dream in which completely illogical, if not absurd events occur that during dreaming seem perfectly natural. Nio's best works are exquisite architectural cadavres in the built environment, which are best appreciated when the viewer is prepared for a brief suspension of disbelief, if it is not surprising that they are located there. Who is willing to suspend his opinion will find works with a sometimes slightly disturbing nature, similar to objects and sculptures by Louise Bourgeois, at once familiar and inexplicable. Beautiful or ugly are words that are rarely relevant.

Bus Stop- "The Amazing Whale Jaw"

A similar dream-like character can be found in independent projects, whether videos or architectural projects without specific assignment. The Fire Emperor belongs to the latter category, this is a project for a market hall in Rotterdam, the presentation of which consists of a sequence of images as a graphic novel. The evocative images of the Fire Emperor are accompanied by equally powerful words: "We want to introduce the Netherlands to the Fire Emperor, a new market being that has adopted the shape of a building. The Fire Emperor will be the hub of public life. Nobody can escape from this voracious building that day and night devours anything that comes near: innocent tourists and experienced gluttons, pale potatoes and fresh coriander, tame pigeons and live squid, dirty market stalls and exclusive restaurants, worn-out musicians and erotic services, rotten smell of durians and the faded perfume of waitresses. Everything is consumed, propelled and ejected again, but not before it changes into a substantially different form under high pressure. "It is not a concrete building, but the suggestion of a possible reality, as the words also express. It is, in other words, a fiction, and the more fictional the work is, the further it is removed from everyday reality, and the stronger it usually is.

The Fire Emperor

The architecture of Nio is stamped with a paraphrase of the author Ben Green's dictum "language + imagination = fiction". For Nio: language + image + imagination = fiction. For him, words and pictures are at the same level, and that Nio is the author of more than one book, emphasizes this. Word and image, articulation and imagination are twin concepts that are close together for Nio as ways to create fiction. Language and image are no resources for him to really describe reality, but to change reality. In that respect, the architecture with little other purpose than to meet practical needs and wishes, in the least interesting. Rather it is the architecture that required little more than being there, that is much more significance. That A house has a certain architectural content, is often taken for granted. If the architecture emerges in unlikely places, it has a greater impact.

The intimate relationship between image and imagination highlights that every architectural proposal is a fiction, for which a suspension of disbelief is required to believe that it will ever become reality. The special feature of the work of Nio is that sometimes it is able to maintain that fictional character. Normally there is poetry in an architectural design that leads to a prosaic result, a building allows itself to be used as a tool. Sometimes, however, the architecture manages to preserve the poetry, and fictional characters is preserved. Maurice Nio has the talent for that. He builds fictions.

Maurice Nio最耐人寻味的作品通常坐落在空旷的基地上。巴士站、高架桥、购物中心和垃圾处理厂周边往往没有太多的建筑，Nio的作品表明，这种环境也可以体现出建筑的丰富内涵。

Nio将普通场所进行了美学激化，产生了一种使人敬而远之的效果，特别是在你不对它抱有期望的情况下更是如此，正如荒谬的事情使得梦境显得完全不合逻辑一样。Nio最好的作品是建成环境中精美的建筑残骸，如果参观者在面对Nio的作品时不感到奇怪的话，他们就要为怀疑的终结而欣喜了。被打断思绪的人们会发现这些作品有点类似Louise Bourgeois的雕塑作品，一旦被人熟知或无法解读就会轻微引起人的不安。我们很难用美丽的或丑陋的词汇来形容他的作品。

Nio其他的个人作品中也有类似的梦幻般的特质，无论是视频作品还是没有特定功能的建筑项目。热情的皇帝就属于后者，这是一个位于鹿特丹的市场大厅，建筑外观由连续的图案组成，看上去就像一部绘画小说。该建筑的设计灵感来源于一段充满力量的文字："我们想将热情的皇帝引入荷兰。这个市场能够适应建筑物的形状，成为公共生活的中心。这幢建筑夜以继日地吞噬着靠近它的任何东西，没有人可以从这个贪吃的建筑中逃脱：无辜的游客、经验丰富的食客、灰白色的土豆和新鲜的香菜、被驯服的鸽子和鲜活的鱿鱼、肮脏的大排档和独立餐馆、穿着破旧的音乐家和色情服务女郎、烂臭的榴莲和女服务员身上散发着微弱味道的香水，一切事物在高压下都会变质，但在那之前都要被一次次地消耗、转化和逐出。"正如这段话所传递的信息那样，这不是一个具体的建筑，而是一个可能实现的建议。换句话说，一部小说的情节越脱离现实，书中描述的未来距离现实越遥远，这部作品就越有力度。

Nio的建筑解释了作家Ben Green的名言"语言+想象力=小说"，而对于Nio则是：语言+图像+想象力=小说。Nio曾不止一次在书中强调：文字和图像是等同的。对于Nio来说，文字和图像，清晰度和想象力是创造小说时关系紧密的两对概念。语言和图像不是用来描述现实的，而是用来改变现实的。在这方面，建筑除了满足实际需要以外，几乎没有其他意义。在一定程度上，如果建筑不仅仅是现实存在的话，它的重要性就会大幅度提升了。一座满足功能需要的房子往往被认为是理所应当的；如果建筑出现在不可能出现的地方，那么它的影响力就会大很多。

形象与想象之间的紧密关系，强调每一个建筑的构思都是虚构的，要想终结怀疑就必须将其付诸实践。Nio作品的特点在于有时它能够保持这种虚构的特性。建筑设计中的诗性往往会导致平淡无奇的设计结果，这使得建筑物本身成为了一种工具。然而有时建筑成功地避免了这种诗性，保留了虚构的特质。Maurice Nio就是这方面的天才，他创造着虚构的建筑。

Beyond Architecture
How to Create Amazing Infrastructure
超越建筑
如何创建令人惊奇的基础设施

The Amazing Whale Jaw
The Aquarians
I'm No Angel
Kiss the Bride
The Monkfish and the Waterwolf
Moon Knight
Prayer of Shadow Protection
Touch of Evil
X-Men

THE AMAZING WHALE JAW

Bus Station in Hoofddorp

Address: Voorplein Spaarne Ziekenhuis, Hoofddorp, Holland
Design: NIO Architecten
Client: Schiphol Project Consult
Contractor: Ooms Bouwmaatschappij
Structural engineer: Ingenieursbureau Zonneveld & Engiplast
Design team: Henk Bultstra, Mirjam Galjé, Hans Larsen, Maurice Nio, Jaakko van 't Spijker
Start design: 1999
Completion: 2003
Costs: euro 1,000,000

惊奇的鲸颚

霍夫多夫汽车站

Photo credit: Hans Pattist, Kees van der Veer, Radek Brunecky

At the beginning of the year 2003, a bus station was built on the forecourt of Hoofddorp's Spaarne Hospital. This facilities block is located in the middle of a square and is a public area in the form of an island that serves as a junction for the local bus service. The design of this kind of building is generally neutral, but here the aim was to create a strong, individual image that was less austere and generic. Hence, the building was designed in the tradition of Oscar Niemeyer as a cross between white modernism and black Baroque.

The building is completely made of polystyrene foam and polyester and is, as such, the world's largest structure in synthetic materials (50m x 10m x 5m). The available budget meant that it could never have been created using conventional construction methods.

People often wonder about the building's shape and what it represents, and there are a number of possible answers. A correct answer in architectural terms is that it can be viewed as a large boulder that has been worn away by footsteps and sight lines. A correct answer in philosophical terms is that it can be regarded as a form that has not been pored over but which simply allowed itself to be discovered. A correct answer in terms of the designing process is that it can also be explained as a product of this process, which in this case was somewhat intuitive because of the unfamiliar technical terrain in which everyone had to operate. All these answers are simultaneously correct and irrelevant. Like the white face of a geisha, every opinion and image can be projected onto the building and it has no answers of its own.

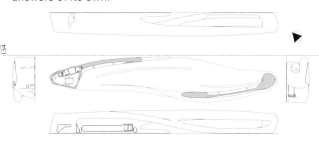

2003年初，在Hoofddorp的Spaarne医院前新建了一座公交站。这个车站位于广场的中间，以岛的形式连接当地的公交服务设施。这种类型的建筑一般会采用较为通用的手法进行设计，但该项目却创造了一个更富个性的有力形象。它的设计延续了奥斯卡·尼迈耶的设计手法，介于白色现代主义和黑色巴洛克风格之间。

该建筑由聚苯乙烯泡沫和聚酯纤维制成，是世界上最大的由合成材料制成的构筑物，体量达到50米×10米×6米。有限的预算意味着这座建筑不可能用传统的施工方法建造。

人们总是询问该建筑的造型是如何形成的，有什么含义。这个问题有很多答案。从建筑的视角看，它可以被看成一块被人们的视线和脚步所侵蚀的巨石。从哲学的视角看，它可以被看成一个尚未被解读但有待被挖掘的形式。从设计过程的角度来看，它也可以作为建造过程的结果，虽然在整个过程中我们由于对技术细节不够熟悉而融入了个人的直觉。所有这些答案都是正确的，它们之间彼此并不相关。这座建筑就像一张艺妓白色的脸，各种不同意见都可以投射到建筑上面，而建筑本身没有自己的答案。

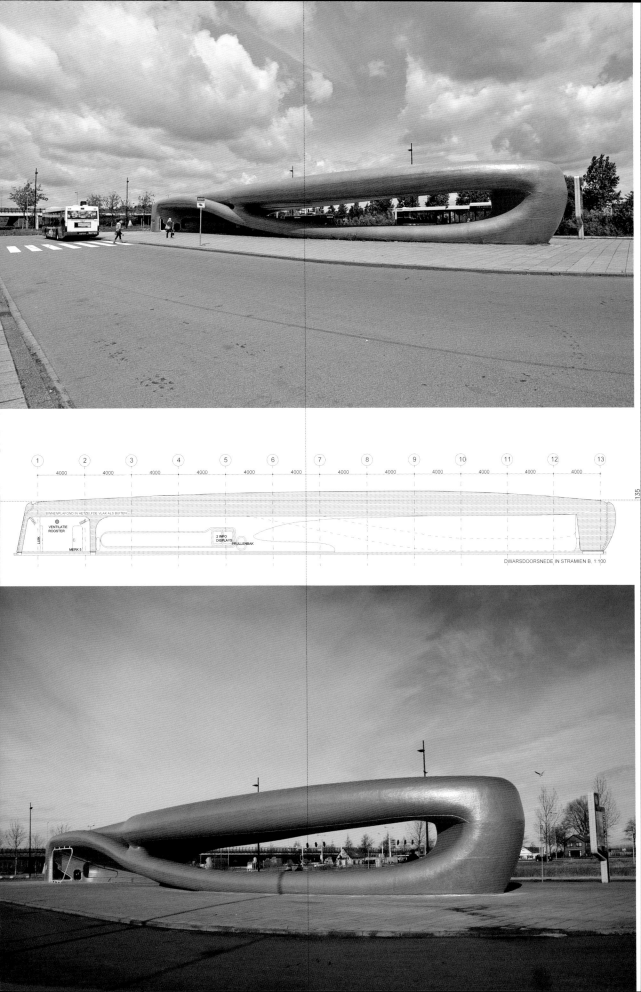

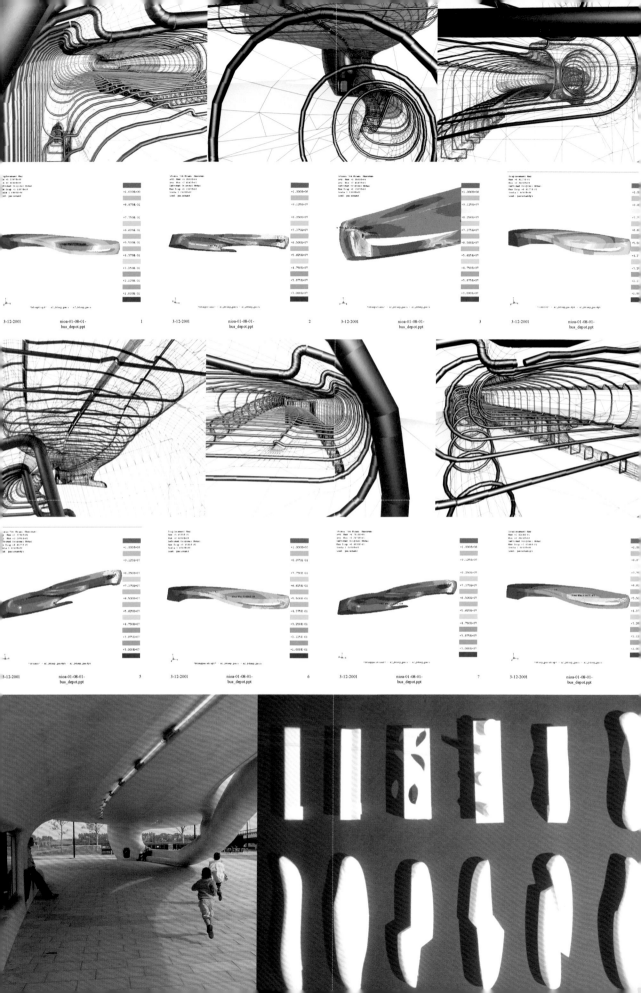

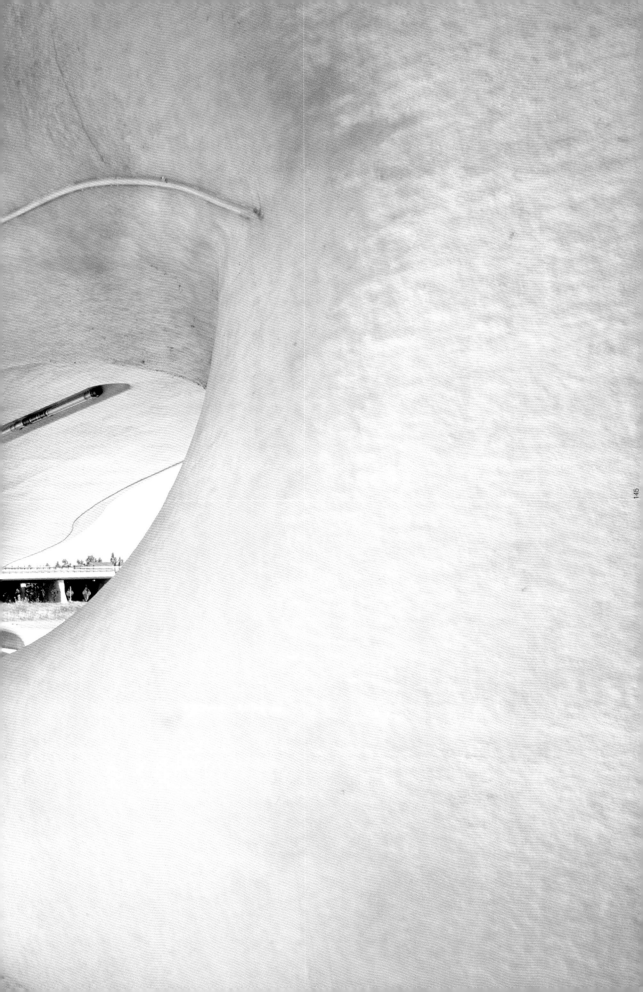

THE AQUARIANS

22 Bridges in the Watertuinen in 's - Hertogenbosch

Address: De Watertuinen, 's-Hertogenbosch, Holland
Design: NIO Architecten
Client: municipality of 's-Hertogenbosch
Contractor: Ippel, Nieuwendijk
Structural engineer: DHV Ruimte en Mobiliteit
Design team: Joan Almekinders, Radek Brunecky, Maurice Nio, Alexander Paschaloudis
Start design: 2004
Completion: 2006
Costs: euro 2,626,000

水瓶座

斯海尔托亨博思水上公园住宅的22座桥

Photo credit: Hans Pattist, Radek Brunecky

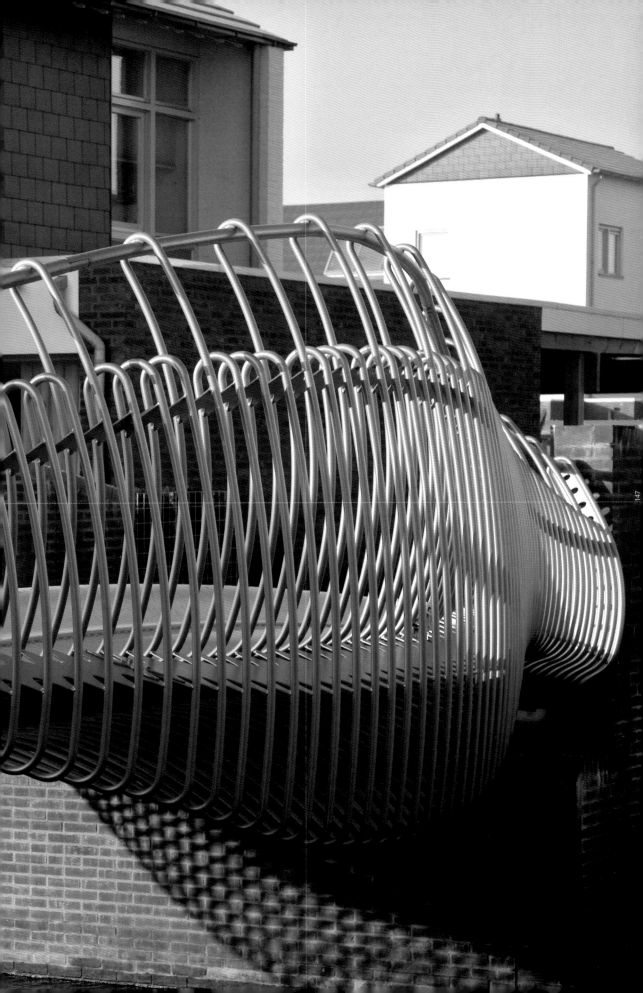

One can easily say that, in spite of the Delta works, the power and force of the water have only increased in the Netherlands. It has mainly adjusted its tactics. Instead of brute force, she now throws in all her charms and is slowly finding her way to areas far outside the polders, to the highest and driest places in the country. Following the slogan of "living by the water" she takes great pleasure in flowing elegantly into new residential areas and is starting to seem unstoppable. This is also the case for Den Bosch, in the new residential area of De Grote Wielen, where waterways were built before roads.

It can hardly, therefore, be called a miracle that the 22 bridges in Den Bosch feel attracted to the water. Infrastructure constructions such as bridges and tunnels have always courageously resisted gravity and water pressure. It has been of no use, the water has already arrived in Brabant. This time no unbending bravery but graceful cooperation instead. If you can't beat them... Just like Bambi seeing her facial traits for the first time in the reflection of a lake, these bridges beam with pure delight. Who would ever have thought of that, having the ideal of beauty at arm's length!

The 22 drops we dribbled on the map of the Watertuinen (Water Gardens) have completely soaked the steel and concrete in the bridges. The apparently stiff and hard materials are completely embraced by the surface tension and now they are wrenching and bending their teardrop form way from one bank to another quay. The bridge is at its widest in the middle, where the steelwork is at its highest and the shape at its fullest. This is the point from where the bridge is no longer a bridge. The point at which passing cars start feeling jealous of the water too.

尽管三角洲对人类有所帮助，但只有在荷兰这个国度，水的力量才算得到了增强。荷兰的河流远离低地，不再使用蛮力，在这个国家海拔最高、最干旱的地方展现出自身的魅力。人们要"依赖水源生活"，河流优雅地注入那些起初不适宜生存的新住宅区。在Den Bosch就是这样。在新住宅区De Grote Wielen，水道的修建甚至要早于公路。

我们不能说Den Bosch的22座桥的奇迹对河水有吸引力。桥梁、隧道等基础设施建设一直在努力抵抗重力和水的压力。但这一点在Brabant并不成立，因为这里的水道是先于公路修建的。该项目用水和桥优雅的合作代替了桥对水的英勇抵抗。如果你不能击败它们……就像小鹿斑比第一次在湖水的倒影中看到自己一样，这些桥传递着喜悦。谁也不会想到的是，这与美的理想还有一大段距离！

我们在Watertuinen（水上公园）地图上滴下的22个水滴都位于钢桥和混凝土桥上。这些坚硬的材料完全被水的表面张力包围着，它们以水滴的方式从堤岸向另一边的码头弯曲。桥的中间宽度最大，那里也是钢结构的最高点和形体最为饱满的地方。在这里，桥已经冲破了传统的概念，路过的车都要开始嫉妒水了。

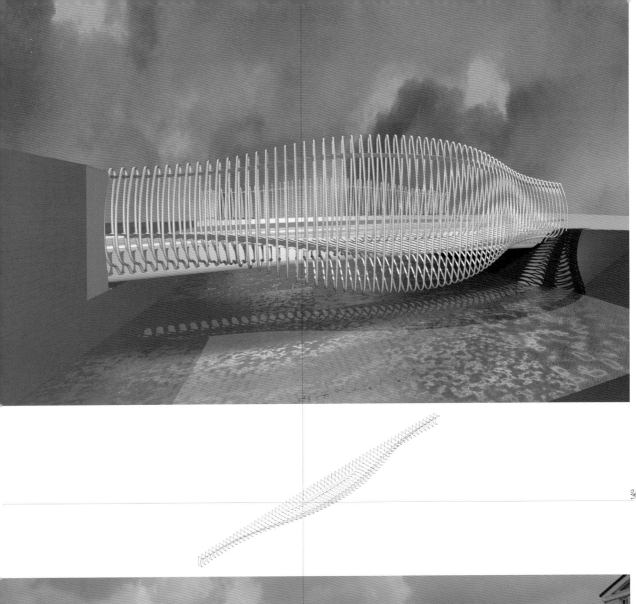

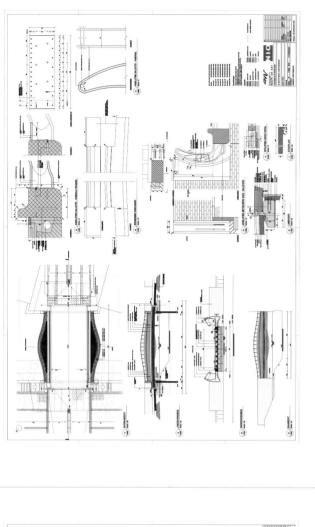
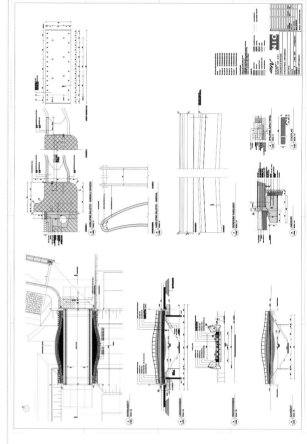
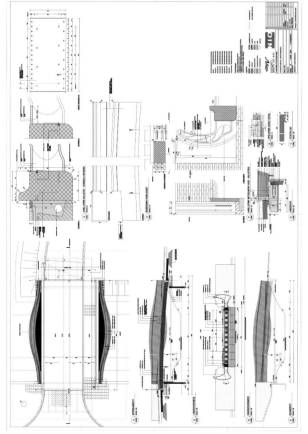
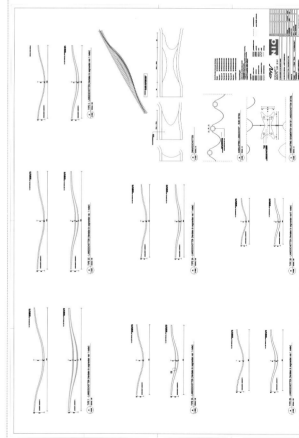

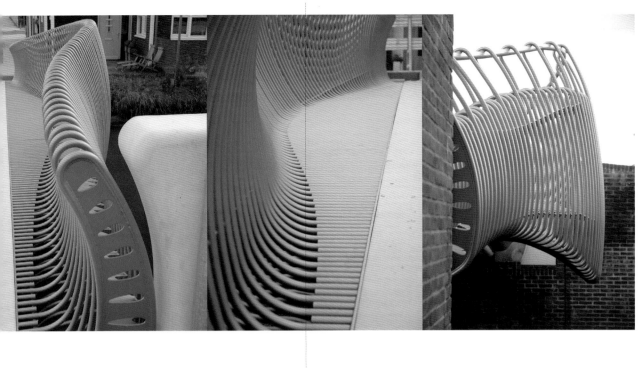
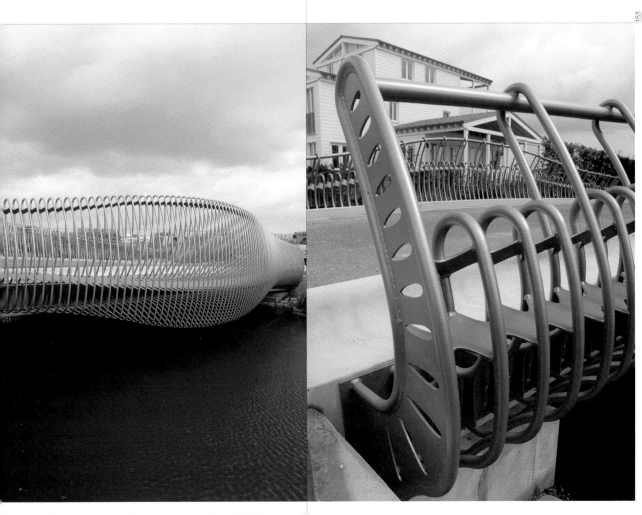

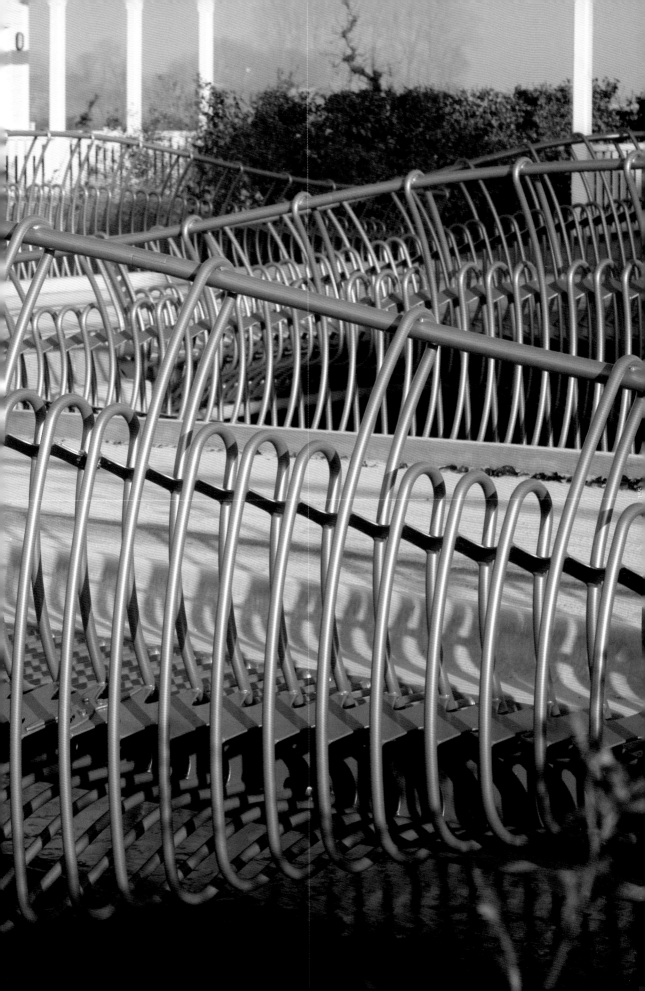

I'M NO ANGEL

Spoortunnel Station Vathorst, Amersfoort

Address: Vathorst, Amersfoort, Holland
Design: NIO Architecten
Client: Ontwikkelingsbedrijf Vathorst
Contractor: Combinatie Kunstwerken Vathorst (Dura Vermeer, BAM Civiel en Heijmans)
Manufacturer: DHV
Design team: Maurice Nio, Arek Seredyn
Start design: 2003
Completion: 2005
Costs: euro 8,600,000
Photo credit: Hans Pattist

我不是天使

阿默斯福特的瓦特豪斯特的火车隧道站

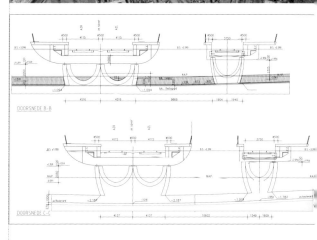

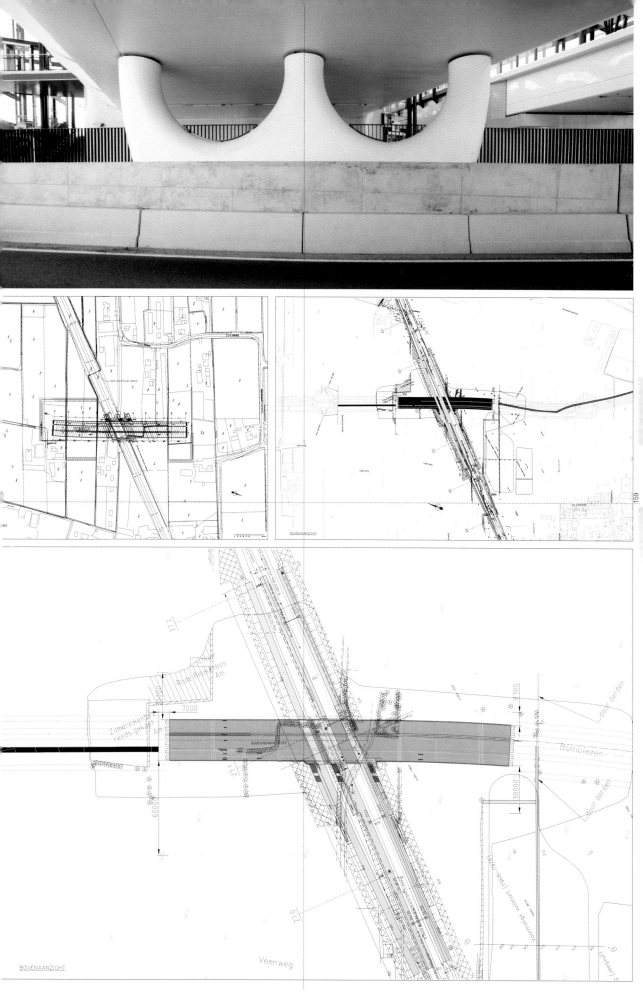

BOVENAANZICHT

BOVENAANZICHT

KISS THE BRIDE

Bridge for bycicles and pedestrians in Zoetermeer

Address: Driemanspolder, Zoetermeer, Holland
Design: NIO Architecten with OKRA landschapsarchitecten
Client: municipality of Zoetermeer
Subcontractor steel: Copier Staalconstructies
Design team: Joan Almekinders, Radek Alexander Hertel
Start design: 2009
Costs: euro 3,950,000

祖特梅尔的自行车和行人过街天桥

亲吻新娘

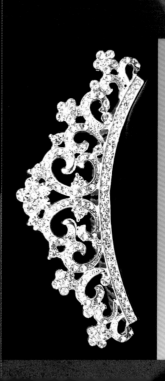

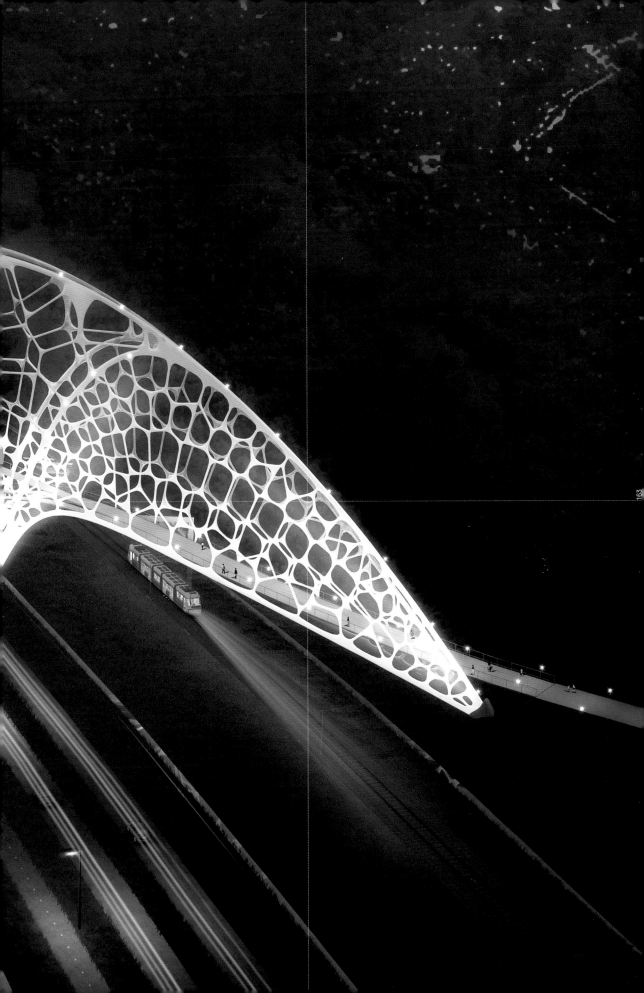

THE MONKFISH AND THE WATERWOLF

A4 Burgerveen-Leiden

Address: the city of Leiden, the city of Leiderdorp, the city of Alkemade, the city of Haarlemmermeer, Holland
Design: NIO Architecten
Client: BAM
Building contractor: BAM
Structural engineer: BAM
Design team: Joan Almekinders, Radek Brunecky, Jan Moritz, Maurice Nio, Giuseppe Vultaggio
Start design: 2006
Completion: 2009
Building costs: euro 250,000,000

安康鱼和水狼

莱顿Burgerveen的A4过街桥

When human intervention takes on the proportion of a natural phenomenon, the question arises whether the nature gods should not be conciliated in advance. What in oriental cultures is not even considered as a form of religion anymore, but as part of everyday life, seems to have disappeared completely from the Dutch culture. In the past, sailors and fishermen were capable of attributing special forces to storms, waters and sea creatures. Nowadays, every kind of involvement in nature is merely seen as a political act while it could just as well be that we have to beware of upsetting the nature gods, especially when we do so much harm to our environment.

In order to curb the daily traffic disaster between Leiden and Burgerveen, the A4 highway will be broadened with one lane in both directions over a distance of 14 kilometres. This means that all the civil works on this route will have to broadened. In Leiden was decided to take advantage of this project by placing the broadened highway immediately in a deepened vessel in order to reduce the noise pollution a much as possible. The lowest point of this civil work is the crossing with the Oude Rijn river, where the cars will pass underneath from now on. The second striking point on this route is an aqueduct as well, on the crossing with the Ringvaart, the border of the Haarlemmermeer, right next to the existing aqueduct – the one with the chessboard.

The ploughing and the drilling of the earth, which is aimed at directing a stinking stream of cars through it every day, requires at least something respectful in return, a place where the Monkfish and the Waterwolf, the Kami (nature deity) of the Oude Rijn and of the Ringvaart, are respected. The Monkfish lies motionless in the sand and feeds itself with whatever he sucks in when he opens his enormous mouth. The Waterwolf, on his turn, used to swallow up entire villages before the Haarlemmermeer was drained in 1852 and now lies as in a century-long hibernation curled up in the Ringdijk. Both aqueducts are the shrines of these two greedy types, a tribute to both monsters. But of course, it does not guarantee a traffic safe passageway.

大自然是否会在人类干预自然的时候事先调解这一矛盾呢？这是一个引人关注的问题。在东方文化中，自然的调节甚至不是一种宗教形式，而是日常生活的一部分，尽管在荷兰文化中这种观点似乎已经完全消失了。过去的船员和渔民们认为有一种特殊的力量在控制风暴、水域和海洋生物。如今我们把人类对海洋的干预看成一种政治行为，但我们必须提防某些行为会惹恼大自然，尤其是那些破坏环境的行为。

为了减少Leiden和Burgerveen两地之间的交通事故，当地政府决定为A4高速公路加建 段14公里长的双向车道。这就意味着整条线路上的所有土建工程都要加建。在Leiden，我们决定将这个项目放在一段加深的隧道中以减少噪音污染。这个工段的最低点是与Oude Rijn河的交点，在这里，汽车将从地下穿越过去。整条路线的第二个重点是要在Haarlemmermeer边缘修建一条临近现有的输水管道的引水渠。

该项目的建设，既是为了管理每天路过此地的大批车流，也是为了表示对安康鱼和水狼（Oude Rijn 和 Ringvaart的自然神）的尊敬。安康鱼在沙堆里一动不动，它用自己的大嘴袭击别的生物并以此为生。水狼曾于1852年Haarlemmermeer干旱之时吞掉了整个村庄，目前它蜷缩在Ringdijk冬眠了一个世纪之久。现在进行施工的这两条引水渠对两种贪婪的怪物充满敬意，虽然它们无法保证道路的行车安全。

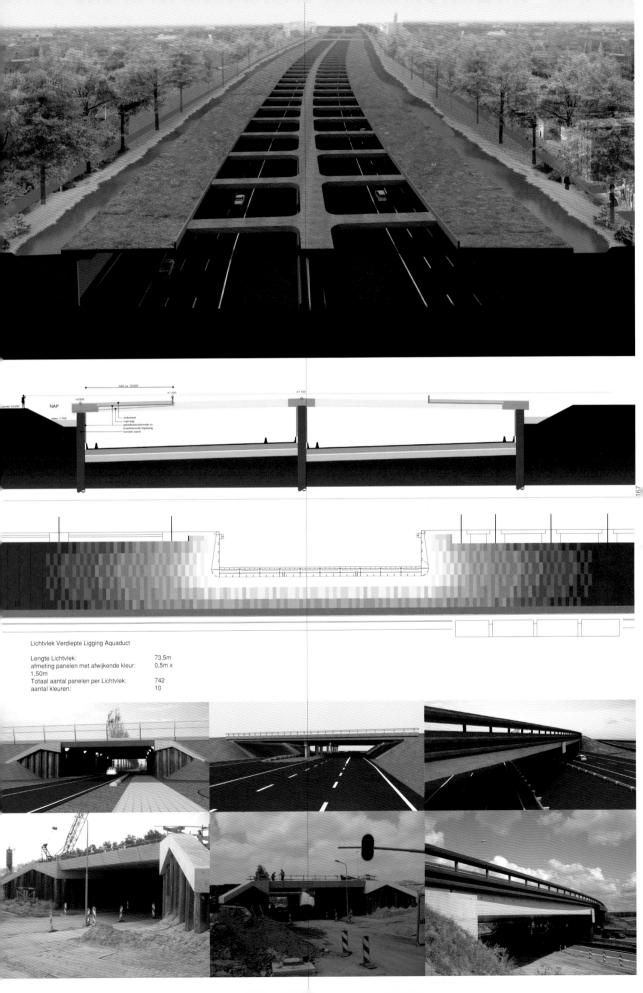

Lichtvlek Verdiepte Ligging Aquaduct

Lengte Lichtvlek: 73,5m
afmeting panelen met afwijkende kleur: 0,5m x 1,50m
Totaal aantal panelen per Lichtvlek: 742
aantal kleuren: 10

MOON KNIGHT

Tunnel Noordammerweg in Amstelveen

Address: Noordammerweg, Amstelveen, Holland
Design: NIO Architecten
Client: Municipality of Amstelveen
Structural engineer: DHV Ruimte en Mobiliteit
Design team: Mark Bitter, Stefano Milani, Maurice Nio
Start design: 2001
Completion: 2004
Building costs: euro 2,600,000

月亮骑士

阿姆斯特尔芬的Noordammerweg隧道

Photo credit: Hans Pattist

As soon as one gets the upper hand, the other one goes underground. It is the same in Amstelveen. Where the Beneluxbaan gets precedence, the other traffic routes that cross it, are built underneath, resulting in tunnels and crossovers. The term "underground" should then be taken both literally and figuratively. These tunnels and crossovers are hidden from the physical space (from the eye) as well as from the mental space (from the concept), dead functional things for which no one could have any feelings. It is an art to make these kinds of spaces specific again and to bring them back to the symbolic circuit.

Is it possible to transform a space that has gone underground and that has been buried as traffic space, into a place to stay? A room that you do not just enter and leave, but that you actually visit? And the other way around: are people capable of behaving like guests instead of hooligans suffering from amnesia, in these kinds of spaces? Not as long as the tunnel only belongs to the municipality, but they will if this is the room of a lucid mind who has decorated the civil construction to his own personal wishes (we suspect, by the way, this is Vasarely's room).

A boy cycles after his own shadow in the tunnel. He lives in Westwijk and commutes every day between his house and his school, which is situated just at the other end of the tunnel. At home he lives in the well-oiled 3-D world of computer games. At school he slowly writes letters and numbers in clumsy little 2-D notebooks. Both worlds are completely incompatible, also in the cycling boy's head. Both worlds are constantly arguing except in this small black-and-white universe in which they embrace (or strangle?) each other and become profoundly 2½-D.

东边日出西边雨，在Amstelveen同样如此。当Beneluxbaan公路得到优先发展的时候，其他与其交叉的交通路线就被改建到了地下。"地下"这个词具有字面和实际两层含义。我们既看不到它们也感受不到它们，在视觉上和思想上都体会不到它们的存在；我们对没有功能的东西都没什么感觉。将这类特定的空间重新变成具有象征意义的线路是一门艺术。

能不能将一个埋入地下的交通空间变成一个可以停留的空间？变成一个人们不仅可以进入和离开，而且还可以访问的房间？其他问题还包括：在这里，人们是否能找到做客的感觉而不是像失忆的街头流氓那样？这段隧道不仅属于当地政府，它还凝结着Vasarely的个人意愿（顺便说一下，我们怀疑这里是Vasarely的房间）。

一个男孩儿在隧道中追着自己的影子骑车。他住在Westwijk，每天骑车上下学，他所就读的学校就在隧道的另一端。男孩儿在家里沉浸于3D电脑游戏的世界，在学校又慢慢地在2D笔记本上书写字母和数字。在单车男孩的意识中，这两个世界完全是不相容的，时刻都在不断地争论。只有在这个小型的黑白空间中，它们才能拥抱（或者扼杀？）在一起，成为一个极度2.5D的世界。

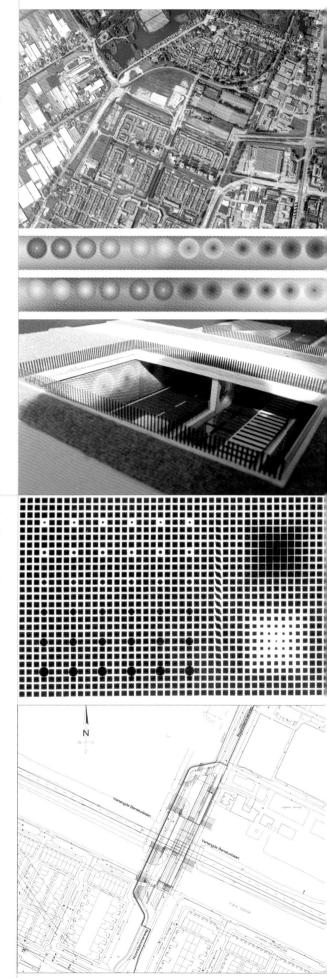

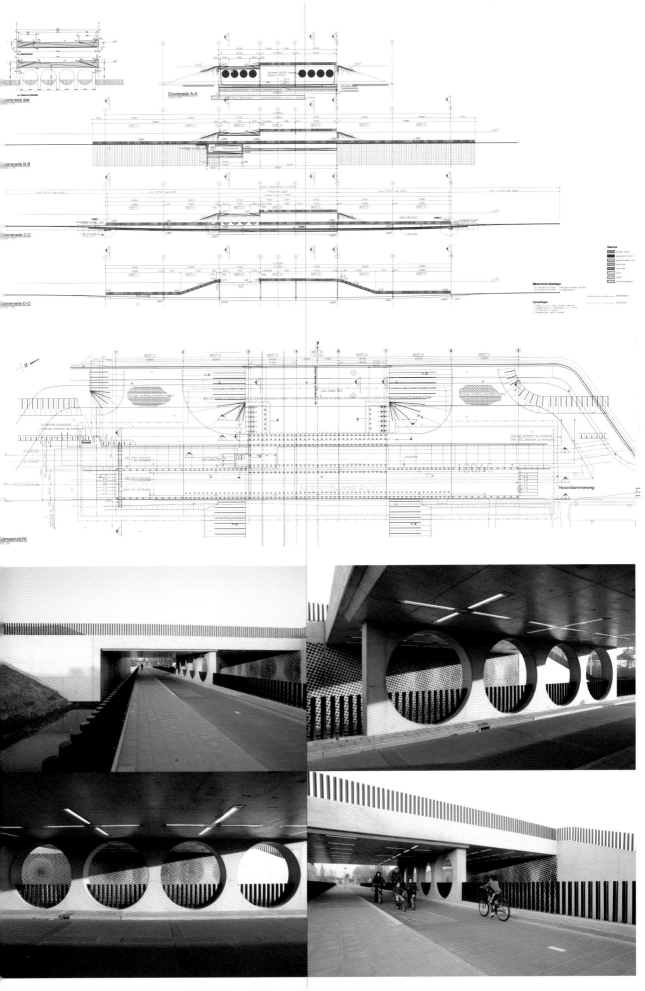

PRAYER OF SHADOW PROTECTION

Bridge Vrouwenakker

Address : Amstel-Drecht canal, Vrouwenakker, the municipality of Liemeer and the municipality of Uithoorn, Holland
Design: NIO Architecten
Client: Province of Noord-Holland
Building contractor: -
Structural engineer: DHV
Design team: Sean Matsumoto, Maurice Nio, Giuseppe V.Jitacgio
Start design: 2006
Completion: 2008
Building costs: euro 6,000,000 (which includes removing the existing bridge and constructing a temporary bridge)

阴影下的祈祷

Vrouwenakker桥

Photo credit: Bianca de Wit, BSB Staalbouw

For most bridges, whether they are fixed or moveable, an architect is not required. They are simply "engineered", as they call it in professional circles, which results in functional constructions devoid of any poetry whatsoever. Unless it be big, prestigious civil works for which the necessary extra financial resources are reserved. That is why it is actually quite nice that for this relatively small bridge an architect was considered necessary, thus escaping civil engineering indifference.

During the design process, we were inspired by the limbs of the praying mantis and the shape of the famous Jedi starfighter from Star Wars, but the language of this bridge corresponds most of all with the design of the pleasure boats that inhabit the Amstel-Drecht canal. It is truly amazing how many different types of these boats there are, every one of them very stylized.

We have made the bridge dark. The colour is something between green and blue, a bit like the water in Holland but more sparkling. The old bridge Vrouwenakker was white, as most bridges are. Why is that? Probably because white is a neutral colour, a colour that does not really express itself. In China white is the colour of death, a reason for us never to choose for it. This bridge is dark, a mysterious and yet elegant figure of which the contours slowly disappear when the evening falls.

在大多数桥梁（无论是固定式桥梁还是可移动式桥梁）的设计过程中，建筑师的参与都不是必不可少的。桥梁设计的专业人士称桥梁是一项简单的"设计"，这种功能性的构筑物缺乏诗性。但当桥的规模较大或建设项目引人注目时，就会有些额外的预算。这就是为什么一座规模不大的桥会请建筑师进行设计的原因，这样可以避免土木工程所带来的冷漠感。

在设计过程中，我们受到了螳螂祈祷时的肢体以及星球大战中绝地战斗机的启发。但整座桥的设计语言还要与Amstel - Drecht运河中的船只相协调。这些船只的类型之多令人惊叹，几乎每只船都代表一种风格。

这座桥的色彩较深，介于绿色和蓝色之间，这种色彩有点像荷兰的水色，但看上去更为闪耀。旧的Vrouwenakker桥是白色的，与大多数桥的色彩一致，也许这是因为白色是中性色，不会真正表达自己。在中国，白色代表死亡，这是我们不使用它的原因之一。这座深色的桥神秘而不失优雅，它的轮廓会慢慢地消失在夜幕中。

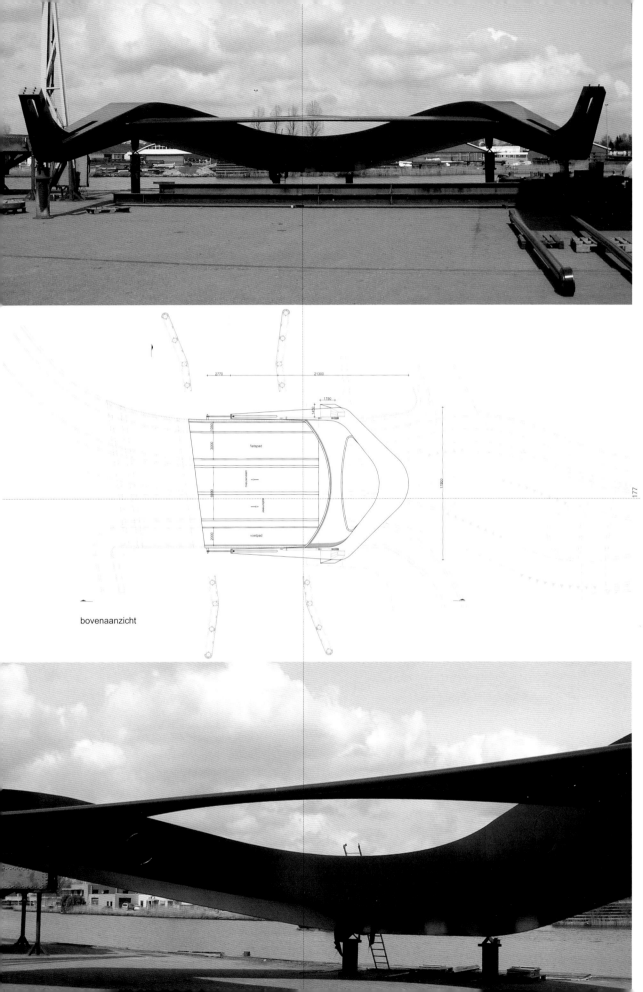

bovenaanzicht

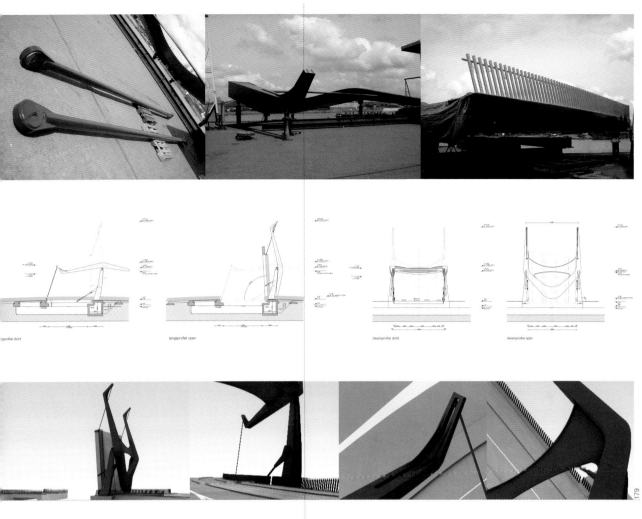

langsprofiel dicht langsprofiel open dwarsprofiel dicht dwarsprofiel open

TOUCH OF EVIL

Interarea tunnel in Pijnacker

Address: VINEX location Tolhek, Pijnacker, Holland
Design: NIO Architecten
Client: Gemeente Pijnacker-Nootdorp
Contractor: Dura Vermeer (Dubbers-Malden)
Structural engineer: DHV/IBZH & Holland Railconsult
Design team: Joan Almekinders, Samir Bantal, Hans Larsen, Enrique Moya-Angeler, Maurice Nio, Arek Seredyn, Tine Silvester-Iversen
Start design: 2002
Completion: 2004
Costs: euro 3,600,000 (10% is spend on architectonical finishing)

邪恶碰触

派纳克的区间隧道

Photo credit: Hans Pattist

What happened here? On the walls and ceilings of the tunnel a strange imprint is visible. A big, whimsical and brightly coloured imprint, an imprint of an unreal and immeasurable shape: as if during the building, an unearthly thing got stuck between the formwork; as if the soul of the former landscape, being rooted up by bulldozers, has gone underground; as if the winding pattern of old polder roads, which has made place for rigid urban developments, takes revenge in the tunnel.

Compared to the furthermore generic tunnel, the thing seems to have loomed up out of nothing. The thing is all the more strange because the rest – the other concrete walls, the asphalt, the paving stones, the fencing, the illumination lines that are at right angles with the road and the bicycle lane/footpath – looks quite normal. The only extravagance being that orange-red thing.

It is not art, even though it was designed and made with the same limitations. It was not invented to externalise an intimate or personal experience. No, on the contrary, it wants to make an inhuman form of life visible in a usually anonymous tunnel. Only in this way, the tunnel is more than a connection between two areas: an alien, immeasurable, asymmetrical and disoriented experience. A tunnel that you will never understand, even though you have been driving through it every day all your life.

这里发生了什么事？在隧道里，我们能看到墙壁和天花板上的奇怪印记。这个巨大的、异想天开的、鲜艳的印记形成了虚幻的、不可名状的形态。这个印记看上去就像一个超越自然的东西卡在了建筑里，又像是从前景观的灵魂被推土机连根拔起，转入了地下，还仿佛是那些古老迂回的道路在隧道里报复城市僵硬的网格。

与同类隧道相比，这个印记似乎是无中生有之物。与周围一切正常的东西（其他混凝土墙、沥青、铺路石、围栏、照明线、车行道、自行车道、行人道）相比，这个印记看上去就更加奇怪了。它是隧道里唯一放纵的橙红色事物。

虽然这一印记的设计和实现是在同样的限制之下完成的，但它并不是艺术品。制作这个印记不是为了将个人的经历具体化，而是要在一个普通的隧道里表现生活的冷漠。只有这样，隧道才不仅仅是两个区域的连接物：它还是一段陌生的、无法估量的、不对称的、迷失方向的生活经历。即使你每天驾驶汽车穿越这个隧道，你也永远不会明白它的意义。

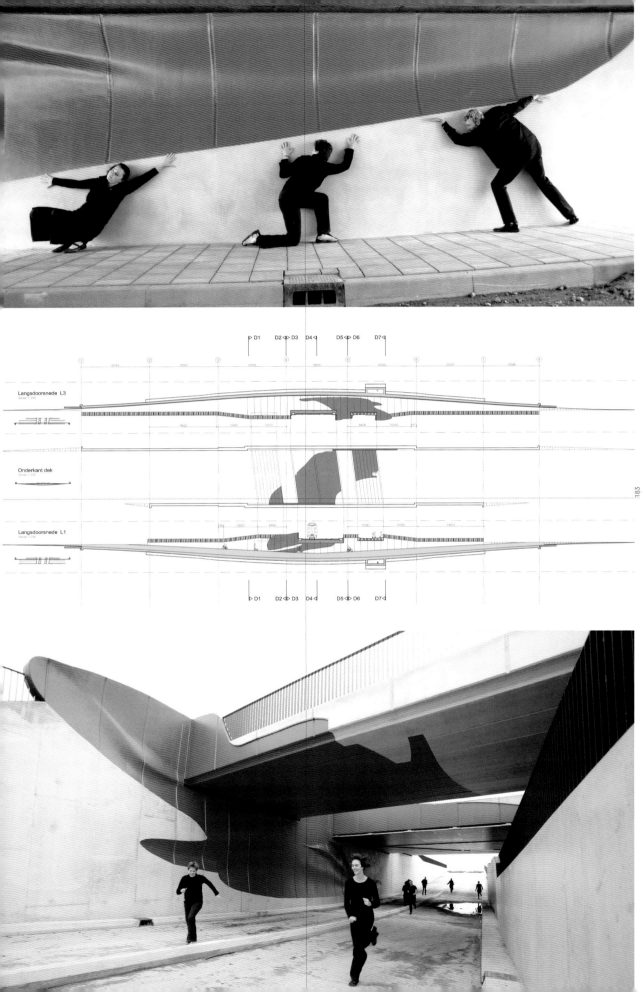

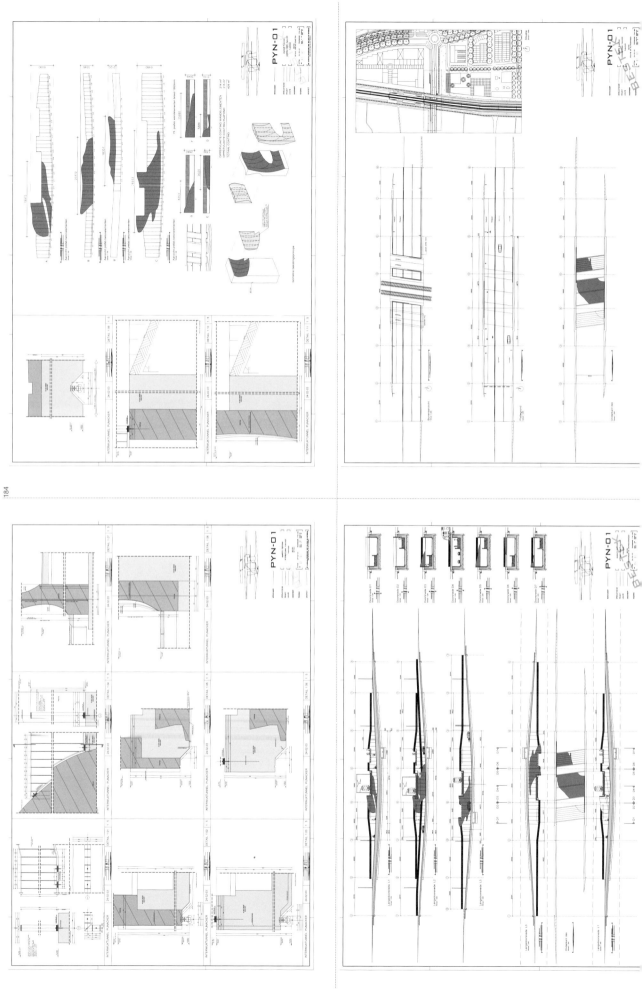

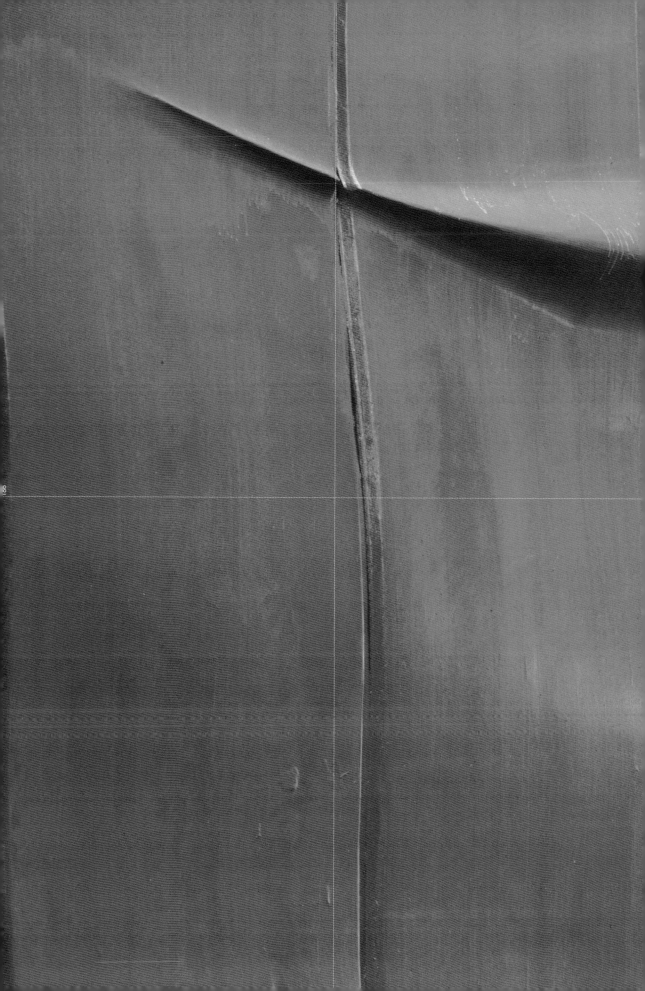

X-MEN

Spoortunnel Vathorst - Noord, Amersfoort

Address: Vathorst, Amersfoort, Holland
Design: NIO Architecten
Client: Ontwikkelingsbedrijf Vathorst
Contractor: Combinatie Kunstwerken Vathorst (Dura Vermeer, BAM Civiel en Heijmans)
Manufacturer: DHV
Design team: Maurice Nio, Remco Arnold
Start design: 2001
Completion: 2005
Costs: euro 8,690,000

X-人

阿默斯福特的瓦特豪斯特的火车隧道

Photo credit: Hans Pattist

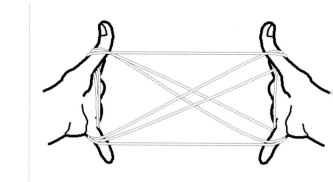

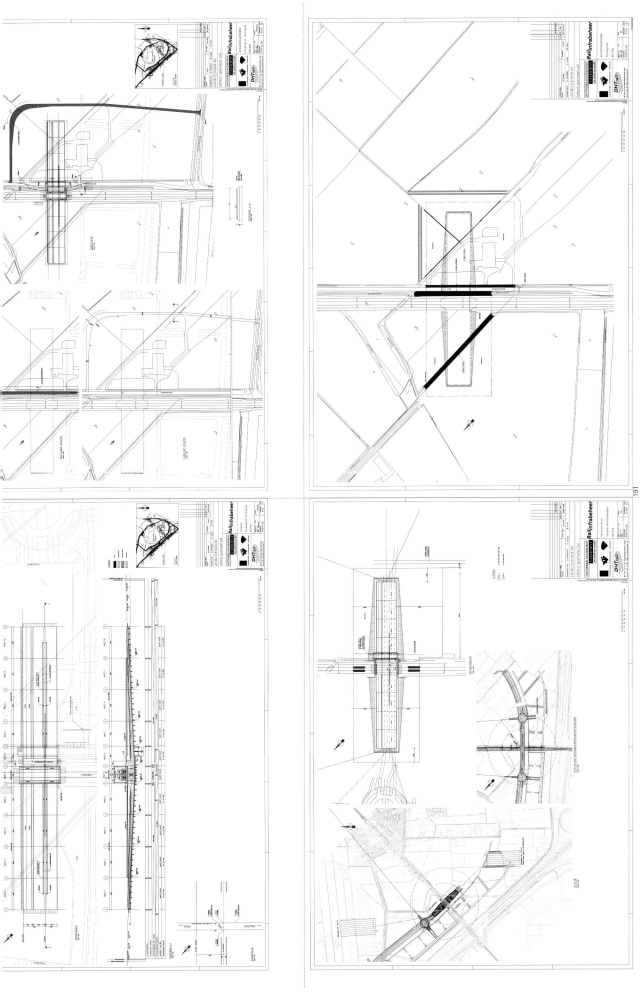

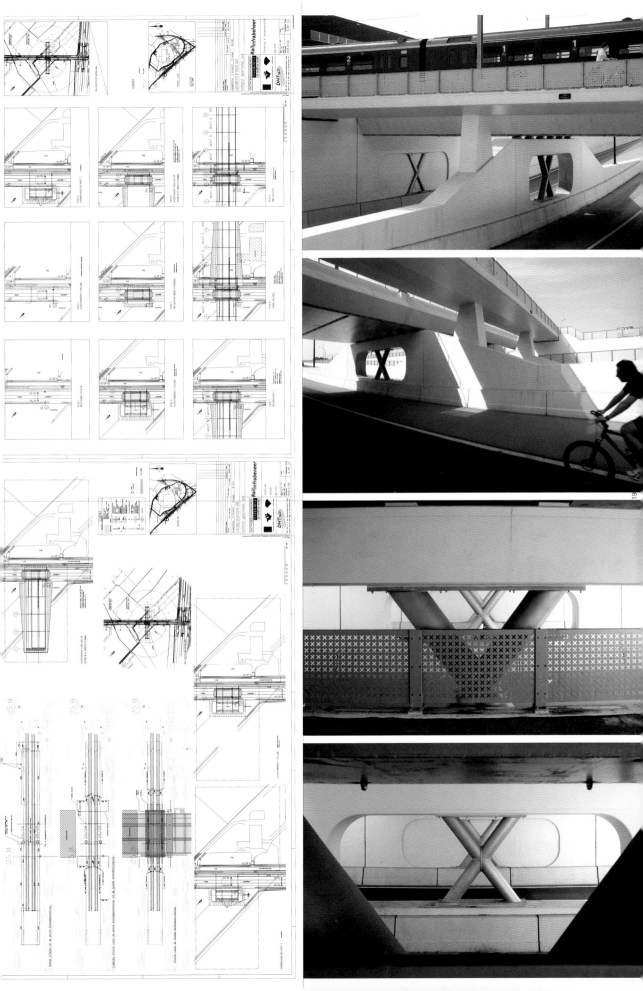

Rethink Urban
New Ways to Think about the City

都市反思
城市新思路

Black Mothafucka
The Eleventh House
Heaven and Hell
House Wide Shut
New Territories
Paradise for Two
Scientia Intuitiva
TwoFace
WWIIOrmhole

BLACK MOTHAFUCKA

Service Building for Wasco in Rotterdam

Address: Sevillaweg / Pittsburghstraat in Noordwest, Rotterdam, Holland
Design: VHP s+a+l
Client: Wasco beheer bv
Contractor: Bouwbedrijf Punte/Systabo Systeembouw
Structural engineer: ATLAS bouwsystemen GmbH, Wesel, Duitsland
Design team: Remco Arnold, Maurice Nio, Jeroen van Oostveen
Start design: 1997
Completion: 1998
Building costs: euro 2,522,000

黑盒子
鹿特丹沃斯科服务大楼

Photo credit: Hans Pattist

Industrial areas retain a remarkable paradox: as each company present there wants to stand out excessively, these zones invariably turn into a monotonous succession of corny exclamation marks. This makes the pale palette stand out, rather than the individual company.

The new industrial zone in Northwest Rotterdam has high quality demands, but does not escape the aforementioned tendency. Due to these circumstances, a new transhipment hall was needed for CV giant WASCO. In order to make the new building stand out in its exuberant non-environment, a non-strategy has been chosen: the program is wrapped in a black armour. Rotterdam is confronted by an indifferent black object, a black mothafucka.

The black box keeps the major part of its mystery stubbornly for itself. It reveals, however, a glimpse of itself in a modest way. Various strategic cuts show something of the logistical process taking place, although each clue is blanketed in a green foggy haze. Lorries drive into the box and unload and load their goods through translucent overhead doors. The black monster absorbs the visitors after they have parked their cars. They simply disappear into the green holes.

工业区存在着一种矛盾：每家公司都希望自己的建筑脱颖而出，这使得工业区充斥着惊叹号（令人惊奇的建筑外观），但这反而突出了那里的苍白，而不是某个公司。

鹿特丹西北部的新工业区虽然有很高的质量要求，但也难逃上述倾向。在这种情况下，CV巨头沃斯科公司要于此兴建一个新的转运厅。为了使新建筑从周边热闹非凡却支离破碎的环境中跳出来，我们采用了无为而治的策略：将建筑包裹在一个黑色的盔甲中。鹿特丹将出现一个冷漠的黑色物体，一个黑色的motha-fucka。

黑盒子保持了建筑本身的神秘感。然而这座建筑一眼望去，却透露出一种谦逊。虽然每一条流线都处于绿色烟雾的笼罩下，建筑形体的切削仍旧体现出一种逻辑过程。卡车穿过半透明的高架门驶入盒子里装卸货物；游客的小汽车驶入建筑后，他们自己也消失在绿色的孔洞里，被黑色的怪物吸收了。

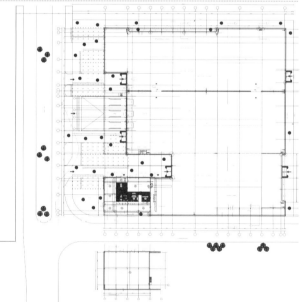

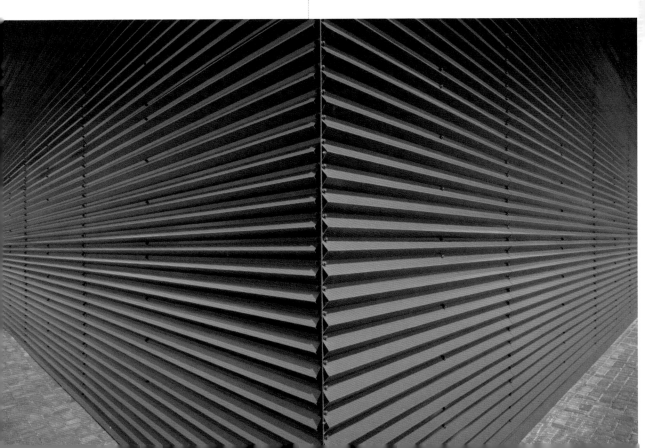

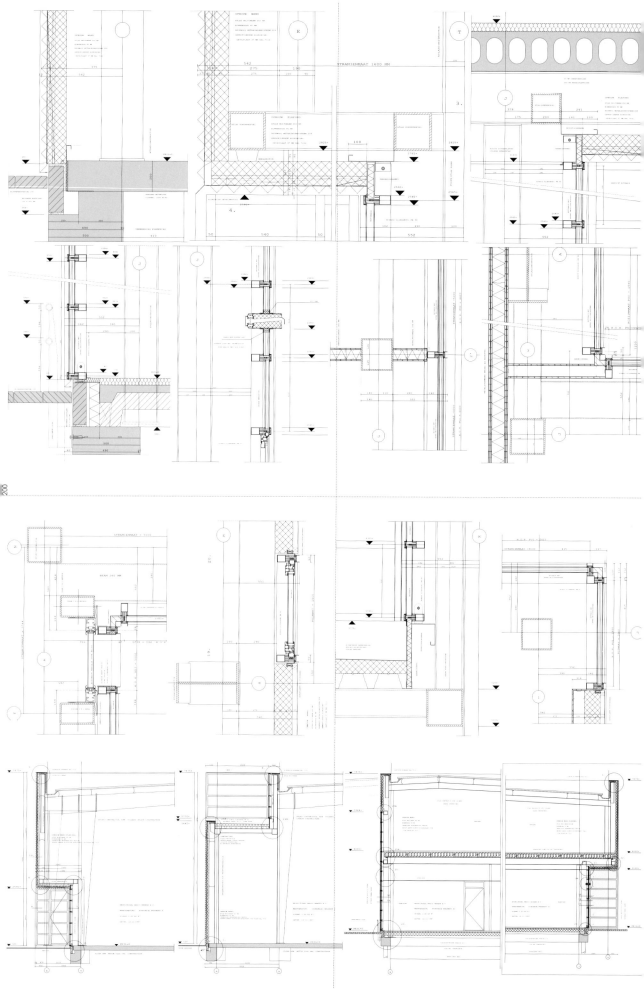

THE ELEVENTH HOUSE

33 Houses in Groenoord-Zuid, Schiedam

Address: Laan van Bol'es, Schiedam, Holland
Design: NIO Architecten
Client: Wijk Ontwikkelingsmaatschappij Groenoord
Contractor: Bouwbedrijf Roos
Structural engineer: Cumae
Design team: Joan Almekinders, Mark Bitter, Hans Larsen, Stefano Milani, Maurice Nio, Sjoerd Roza
Start design: 2001
Completion: 2005
Costs: euro 3,600,000

第十一座住宅

斯希丹Groenoord-Zuid的33座住宅

Photo credit: Hans Pattist

In the middle of an area dominated by the usual porched houses, gallery flats and maisonettes from the 1970s, three identical blocks with a total of 33 ground-bound houses have been designed. Identical however? The residents have the choice to expand their house with an extra room on top of their roof shaped like an eye that overlooks the area. Therefore, the three blocks will never be identical after all. Secret number one.

These three blocks are literally newcomers to the Groenoord-Zuid neighbourhood. They do everything differently. As opposed to the anonymous, green public spaces that surround the flats, in this design was chosen for small, intimate outside spaces that serve as an extension of the houses. As opposed to the classic and strict division between the front façade and the back facade of the flats, these three blocks only have a front. As opposed to the ruthlessly hard roof edges of the flats, is the irregular ridge created by the optional attic rooms and the incisions of the terraces. As opposed to the clearly countable repetition of the houses in the flats, the three blocks have a façade from which it cannot be read whether there are two or twenty houses behind it. Secret number two.

There they are, the blocks, like hollow monoliths on the Laan van Bol'es, like dark trunks in between the green. They reveal nothing on the outside. Everything takes place on the inside. And yet, in each block there is one house that always escapes the direction of introversion. It is the eleventh house. The house with two front doors. The house with an isolated outside storeroom. The house with an average back garden that has become a patio. The house with the most common floor plan compared to the other ten. Secret number three.

我们利用33个绑定在地面上的住宅组成了三个相同的街区，当地住宅以带门廊的独栋住宅、画廊公寓和建造于20世纪70年代的小屋为主。这些房子的相同之处是居民可以在自家屋顶加建一个形如眼睛的房间，俯瞰当地景观。而这三个街区绝不会是雷同的。这是项目中的第一个谜团。

这三个街区毗邻Groenoord-Zuid。它们的一切都不同于旧街区。为了避免街区缺乏个性，我们将公寓周边的绿色公共空间作为建筑的延伸，设计成了小型的宜人室外空间。为了避免公寓正立面和背立面之间严格的划分，这三个街区都只有正立面。为了避免出现屋顶出现硬质边缘，我们用加建的阁楼和台阶切口创造出不规则的屋脊。为了避免公寓的重复性，这三个街区从立面上看无法判断临街建筑后面还有多少房子。这是项目中的第二个谜团。

这三个街区在Laan van Bol'es就像空心的巨型石像、绿色背景中的暗色中继线。从它们的外表上看不出什么，一切玄机都在街区内部。然而，每个街区都有一所房子不具备内向性的特征。那就是第十一座房子。它有两个前门、一个孤立的外部储藏室和一个普通的后花园（已成为露台），与其他十座房子相比，这座房子的平面却是最普通的。这是项目中的第三个谜团。

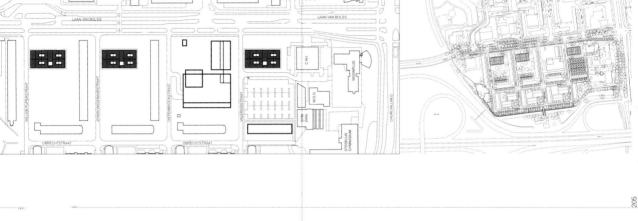

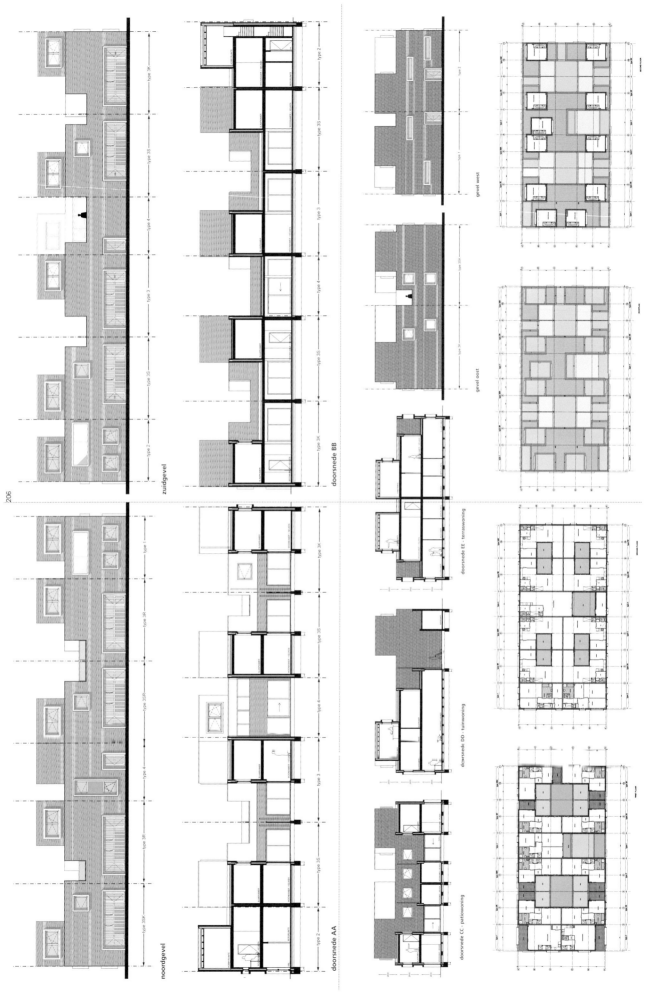

HEAVEN AND HELL

Life Style Department Store in Voorburg

Address: Herenstraat 177, Voorburg, Holland
Design: NIO Architecten
Client: TCN Property Projects/The Village CV
Contractor: Bouw- en Aannemingsbedrijf Dessing/ATOP afbouw/Araform/Xylos meubelmakers
Structural engineer: Ingenieursbureau Zonneveld
Design team: Samir Bantal, Tershia Habbitts, Christel van der Hulst, Fevzi Köstüre, Hans Larsen, Alessandro Mandalà, Maurice Nio, Stephen Pasterkamp, Jaakko van 't Spijker
Start design: 1998
Completion: 2002
Costs: euro 2,500,000

Photo credit: Hans Pattist

天堂与地狱

莱孚丹生活时尚百货商店

In a continuation of the Herenstraat in Voorburg and in the underworld of the Utrechtse Baan the first Life Style Department Store of the Netherlands is realized, a store in which different brands of clothing, shoes, accessories, cosmetics and cappuccino are being sold and that, in contrast with the grey and hard civil world of the viaduct, in its colour scheme and contours evokes the strangeness of a different world.

As soon as you walk through the colourful outer facades of glass mosaic, you will enter yet another world, a low and difficult to define space, a black and white theatre in which the clothes are the most important actors. The white walls, ceilings and display furniture form the side wings that are being carried by a stage of black-pigmented concrete. Only the bar, the couch and some walls in the restaurant are red.

It was already clear from the beginning that the store would not be divided into separate little shops, but that the different brands and articles would be displayed in one big space. It was the proprietor's specific wish to keep the finish and interior of the store space modest and black and white and to carry through only the curved lines of the exterior on the inside, thus arising a store space with furniture lined up along a virtual and pleated thread, a thread that, like in Ariadne's myth, you can follow to be all loaded up outside again.

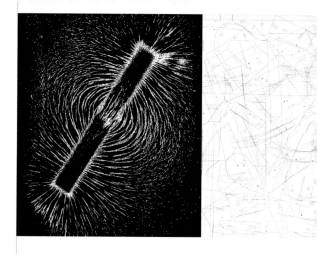

荷兰第一家生活时尚百货商店在Voorburg Herenstraat的延伸地带建成了。商店位于Utrechtse Baan高架桥下面，出售不同品牌的服装、鞋、饰品、化妆品和卡布奇诺。与灰色坚硬的高架桥相比，商店建筑的配色方案和独特的外形令人感觉身处另一个世界。

当你穿过色彩斑斓的玻璃马赛克外墙后，就进入了另一个世界。这是一个低矮而难以界定的空间，在以黑色和白色为主色调的空间中，衣服是最重要的部分。白色的部分包括墙壁、天花板和侧翼所展示的家具（由掺有黑色颜料的混凝土制成）。只有酒吧、餐厅的沙发和一些墙壁是红色的。

从一开始，这家商店就不会被划分成独立的小商店，而是要将不同品牌的商品陈列在一个大空间里。业主特别希望商店内的储藏空间保持低调的黑白色彩，并将室外的弧线元素引入室内。如此产生的储藏空间中，家具沿一条虚拟的折线呈线性排列，就像Ariadne的神话一样，你可以带着所有东西沿着此路线再次走出商店。

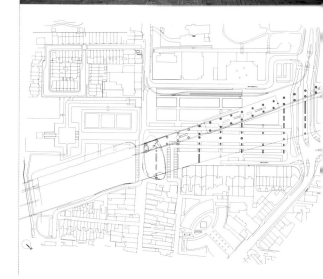

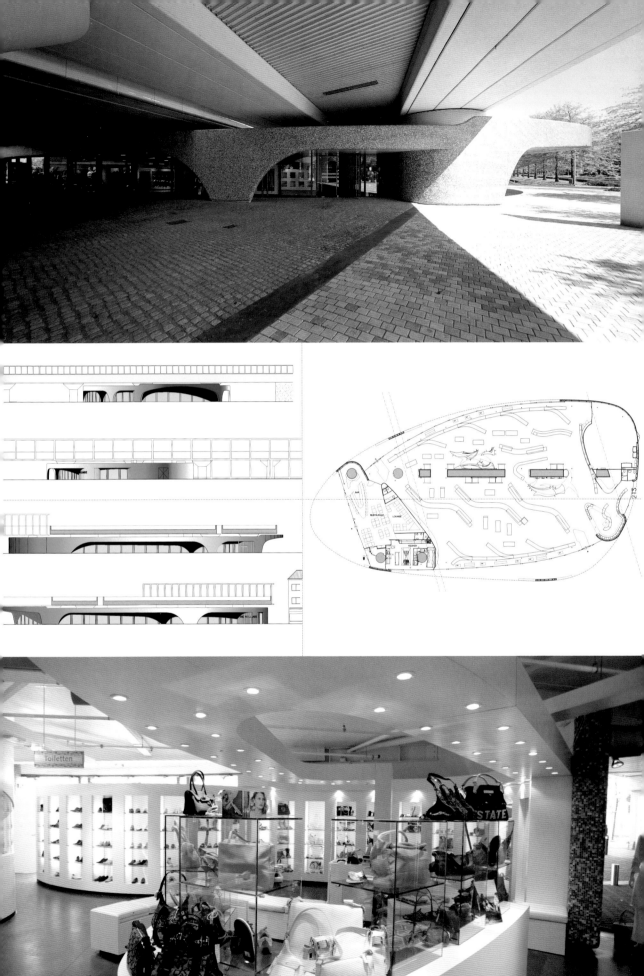

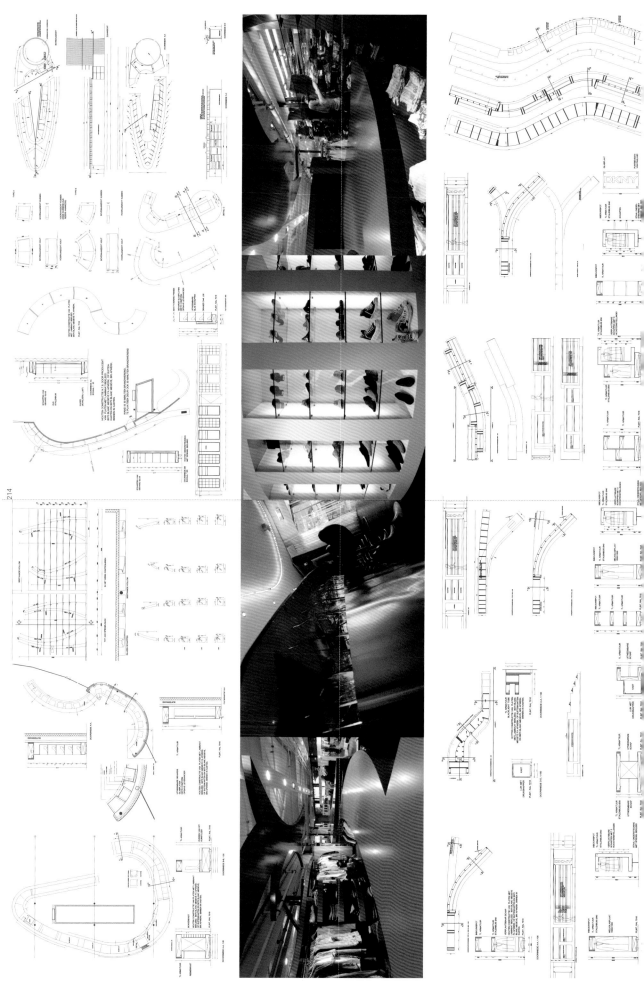

HOUSE WIDE SHUT

48 Groundbound Houses in Westzaan

Address: Westzaner Werf, Westzaan, the Netherlands
Design: NIO Architecten
Client: De Woonmij Zaanstad
Constructor: Peters en Van Leeuwen
Design team: Joost Kok, Stefano Milani, Maurice Nio, Arek Seredyn
Start desing: 2003
Realisation: 2008
Costs: euro 8,500,000

大开口住宅

Westzaan的48栋地景住宅

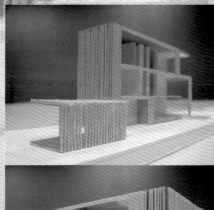

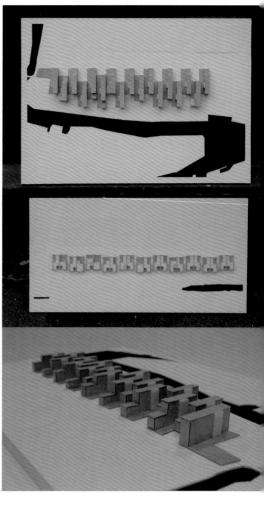
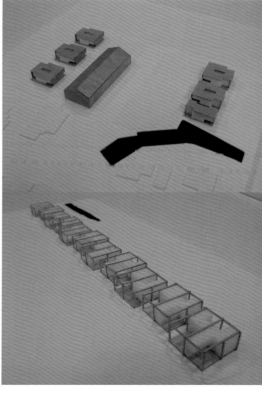

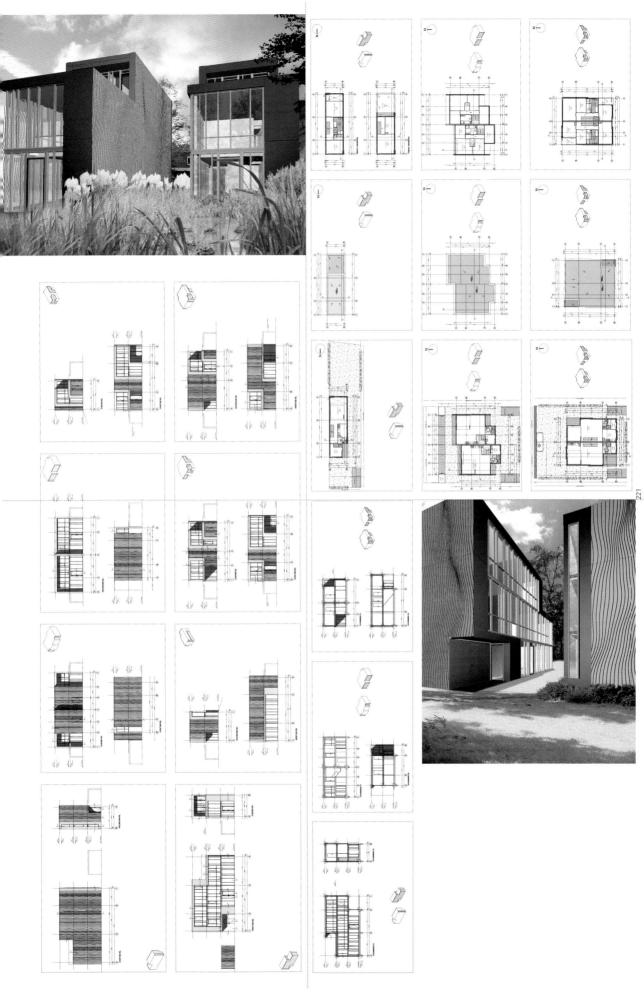

NEW TERRITORIES

Strategy for Renewal of Business Area

Address: Mijdrecht & Strompolder, Rotterdam, Holland
Design: NIO Architecten
Developing partner: RECC
Design team: Joan Almekinders, Sean Matsumoto, Maurice Nio
Start design: 2005

新领域

商务区的重建策略

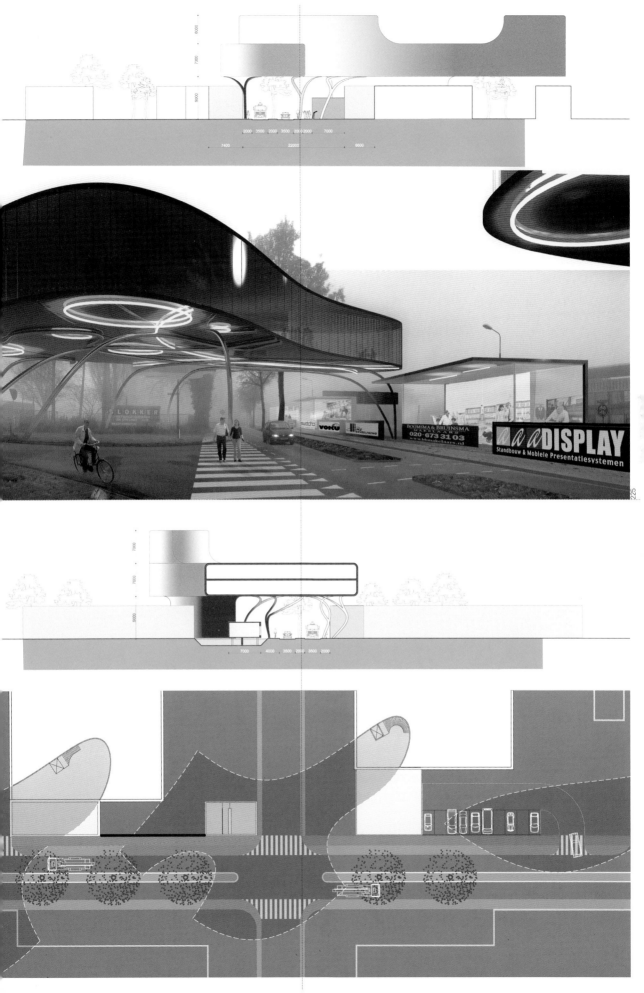

PARADISE FOR TWO

Pavilion and house at the Botshol

Address: Botshol, Holland
Design: NIO Architecten, Rotterdam
Client: Private person
Delegated client and building management: RECC, Vinkeveen
Building contractor: D. Kroese bv, Vinkeveen
Structural engineer: Cumae, Arnhem
Design team: Maurice Nio, Arek Seredyn
Start design: 2003
Completion: 2005 (pavilion) - 2007 (house)
Building costs: euro 220,000 (pavilion)

Photo credit: Hans Pattist

两个人的天堂

Botshol的展馆和住宅

The grass and marshland of the Botshol is a nature reserve and a unique habitat for unusual plants (wild orchids, rare marsh plants, cotton grass, holly-leaved naiad, stonewort, bog spurge) and birds (the buzzard, the harrier, the cormorant, the bittern, the red-crested pochard, the spoonbill). It is situated – a bit hidden – as an oasis in the midst of the Randstad. It is obvious that in the Botshol cannot be built, but also in the surrounding area of De Ronde Venen there are restrictions to new housing development or partial renewal of existing farms. This is why a lot stays the same. It is the art to create something here that answers the current wishes and demands of the inhabitants and at the same time fits the new plans with nature (the direct surroundings of the Botshol have been designated in 2000 as nature development area, as a consequence, the agricultural function slowly disappears to make room for wet marsh nature).

Precisely, where the austere lay-out of the peat grass area touches the new nature and the whimsical landscape of the Botshol, is a garden of which the inner world does not reveal itself. Just like the Botshol, the inside of the garden is hidden and it shows, equally sovereign, only its green edges to the outside world. Almost invisible from outside are two buildings, a farm and an annex, both ready for renewal. In line with the heart of the garden, which has been cultivated and even seems a little heavenly, the renewal was used to intensify the unexpected atmosphere of this location.

The basis of the design are two distinct zinc roofs under which different spaces are given a place in a very relaxed manner. One part is now ready: the annex has been transformed into a pavilion. The zinc roof is cheekily placed diagonally over the square wooden building. The struts of the roof fan out, as if they are trying to imitate the pattern of the nearby Vinkeveense plassen, and the dark stained wooden duckboards are lifted compared to ground level, so that the building seems to float above the garden. The image of the pavilion is new and possibly unexpected, but eventually it is equally natural and graceful as a cormorant in the Botshol.

Botshol有一个草地和沼泽自然保护区，那里是一些珍稀植物（野生兰花、稀有的沼泽植物、棉花草、冬青叶稚虫、轮藻、沼泽大戟）和鸟类（鹫、鹞、鸬鹚、麻鸭、红头潜鸭、琵鹭）的栖息地。保护区位置隐蔽，位于Randstad中部的绿洲中。Botshol禁止修建房屋，在周边的De Ronde Venen地区也禁止修建住房或翻新现有农场。这也是当地面貌保持不变的原因。只有用艺术才能创造出满足当前居民意愿和要求的东西，同时与新的自然计划更好地结合。（Botshol及周边地区于2000年被指定为自然发展区，当地的农业用地慢慢被潮湿的沼泽取代了。）

原始的沼泽地区与Botshol的新景观相结合，形成了一个内向性的花园。这个花园和Botshol一样，将自己的内在隐藏起来，这里只能看到外面世界的绿色边缘。花园内部是等待翻新的农场和一幢附属建筑，但从外面几乎看不到它们。花园的中心有一片精耕细作的田地，重建的建筑渲染了这里的气氛，令人耳目一新。

设计的出发点是要以一种轻松的方式，在两个不同的锌制屋顶下布置两个截然不同的空间。其中的一个已准备就绪：附属建筑已被改造成一座展馆。锌制屋顶以对角线方向贴在方形木建筑屋顶。外露的风机仿佛在模仿plassen的Vinkeveense风格，地面上方有一层深色的褐色木板，整幢建筑物似乎漂浮在花园之上。展馆的新形象虽然令人意外，但它的本质还是像Botshol的鸬鹚一般自然、优雅。

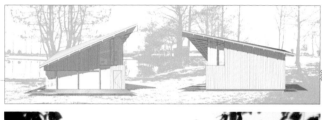

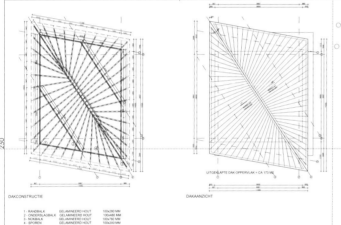

DAKCONSTRUCTIE

DAKAANZICHT

1 - RANDBALK   GELAMINEERD HOUT   100x280 MM
2 - ONDERSLAGBALK   GELAMINEERD HOUT   100x480 MM
3 - NOKBALK   GELAMINEERD HOUT   100x780 MM
4 - SPOREN   GELAMINEERD HOUT   100x200 MM

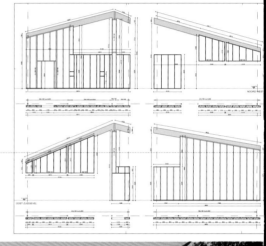

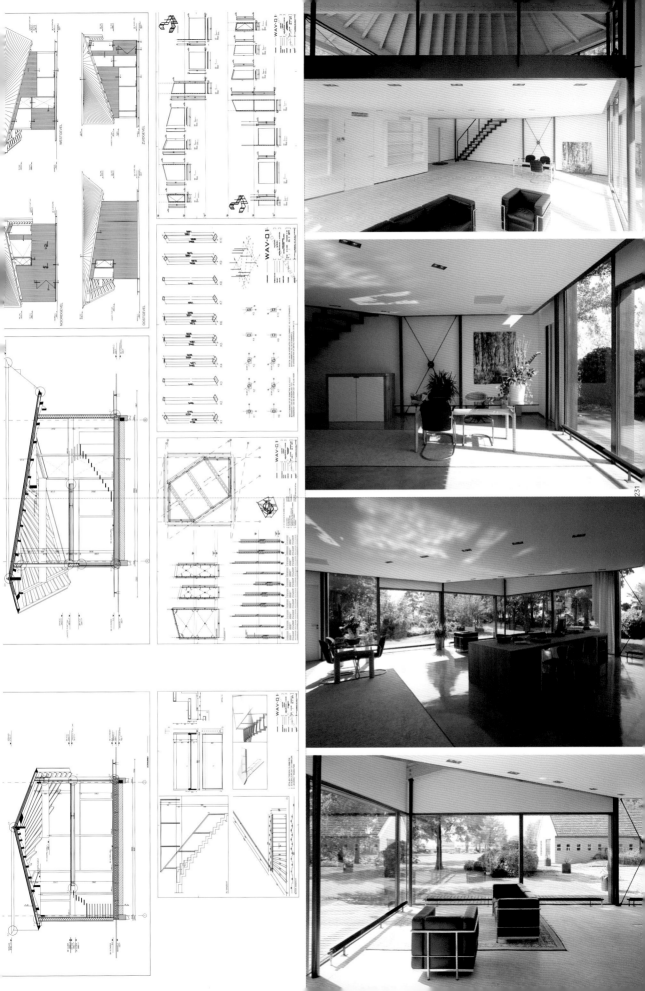

SCIENTIA INTUITIVA

The Spinoza Tower in Amsterdam

Address: Waterlooplein/Mr. Visserplein, Amsterdam, Holland
Design: NIO Architecten
Client: foundation Spinoza Centrum Amsterdam
Design team: Joost Kok, Maurice Nio, Jan Willem Terlouw
Start design: 2008
Completion: 2010

科学直觉
阿姆斯特丹的斯宾诺莎塔

It seems a contradiction: something material as tribute to Spinoza. Would a purely spiritual hymn are not appropriate? The answer is simply "No", as written and spoken hymns of Spinoza there are enough. What we need, what Amsterdam needs is a built gesture that is as sublime as the thinking of the radical verlichter. No statue, but a public facility, a universally accessible center that not only Spinoza in the spotlight but also to Amsterdam as the capital of Spinoza.

The proposal is fully in line with the tower tradition of Amsterdam (Montelbaantoren, Haringpakkerstoren, Schreierstoren, to name but a few) and also embodies the "ladder" of Spinoza. Underneath, on the ground floor space for the first form of knowledge – "opinion" or "imagination" – based on vague, arbitrary experience. The second level is the space for the mind, reason, clear, and systematic welonderscheiden defined. But what matters is the above space, the space for the intuitive know, for the total "transparency" of the essence of things.

In the center of Amsterdam are many places for the tower to Spinoza, but perhaps, it is Mr. Visserplein the most ideal location. Here, near the place where Spinoza was born, is the shiny facade of the tower, three religious institutions reflect the seventeenth-century Portuguese-Israeelitische Synagogue, the High German Synagogue (now part of the Jewish Historical Museum) and the nineteenth-century Moses and Aaron Church . What's better than this direct illustration of pure spinozistische Amsterdam and tolerance?

这看上去很矛盾：斯宾诺莎是一位哲学家，但他得到的都是物质上的奖赏。难道不应该给他一些纯精神的赞歌么？答案当然是"不"，因为斯宾诺莎已经得到了足够多的赞歌。人们所需要的，阿姆斯特丹所需要的，是和斯宾诺莎激进思想一样伟大的姿态。这种姿态不是雕像而是一处公共设施，一个被人们频繁造访的地方。在这里，主角不仅仅是斯宾诺莎，也是斯宾诺莎所在地的首都——阿姆斯特丹。

方案设计遵循阿姆斯特丹传统的造塔理念（如Montelbaantoren塔、Haringpakkerstoren塔和Schreierstoren塔等都是以此理念建造出来的），同时体现斯宾诺莎的"阶梯"思想。在塔的底部，一层空间代表了知识的第一种形式：含混而主观的"观念"或"想象"。二层则代表了思维、理性、清晰性和系统性等概念的分离。更为重要的上层空间则代表着直觉和事物本质的"透明性"。

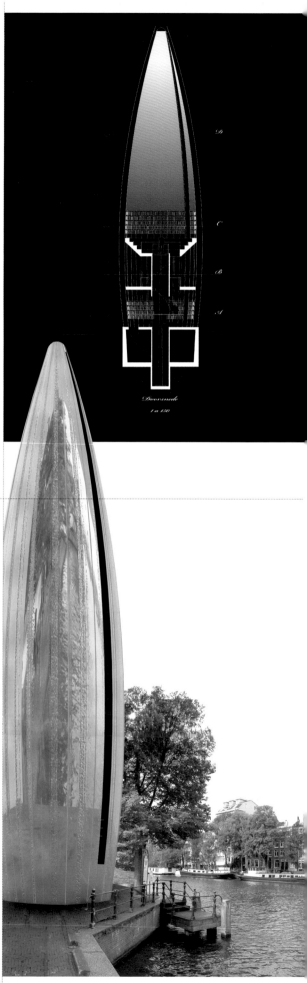

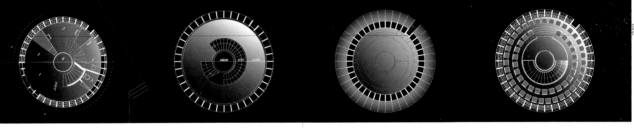

TWOFACE

198 Apartments in 'Tussen de Vaarten' in Almere-Stad

Address: Renoirstraat (Tussen de Vaarten), Almere-Stad, Holland
Design: NIO Architecten
Client: Bouwmaatschappij Verwelius bv, Amsterdam
Building contractor: Bouwmaatschappij Verwelius bv, Amsterdam
Structural engineer: Cumae bv
Design team: Remco Arnold, Mark Bitter, Maurice Nio, Jaakko van 't Spijker
Start design: 2000
Completion: phased over 2003–2004
Building costs: euro 11,000,000 (119 apartments, high-rise building) + euro 8,800,000 (79 houses, low-rise building)

双面人

位于阿尔默勒运河之间的198套公寓

Photo credit: Hans Pattist

With every house-building assignment, it is the art to escape the standard plans that are in the top drawers of estate agents and developers in a relaxed manner. That applies especially to this particular assignment in the "Tussen de Vaarten" area in Almere, where urban developers have come up with a lingering block of houses of 400 metres long on the "Hoge Vaart", consisting of 119 apartments with walkways on the northeast side to open them up. Because the block only consists of three floors, we have proposed to give every apartment its own entrance hall on the ground floor, to prevent walkways and elevators and to minimize the service costs. This paves the way for developing a mosaic of different types of apartments (21 in total) along the entire depth of 15 meters of the nave, for opening up the flats on the different floors and for creating an offer as diverse as possible.

That the apartments all have their own front door on ground level is not the only special feature. What strikes are the 119 collective mini-private-gardens to which every inhabitant has given his highly personal interpretation; the wooden planking of five metres deep which borders on the reed ponds; the absence of the – according to the old building decree – obligatory extra door in the façade in front of the storeroom (119 front doors in a row seemed radical enough to us); the conservatories and patios of the apartments on the floors (has anyone ever made patios in apartment buildings?); the "floating" rooms between the entrance on the ground floor and the living space on the second floor, and, of course, the exceptions to the exceptions: the front houses that usually consist of three floors and of which the biggest one takes up 186 square metres.

Because the urban layout of the area dictated an ecological look, it would have been obvious to cover the façade on the "Grote Vaart" with wood, of which the parts (how could it be any different?) form a structure of open slats. We, however, have come up with a cool milky white glass façade that disguises the typological turmoil and – very ecologically – reflects the reeds and the water. By vertically weatherboarding the glass, the veil-like effect becomes much better visible than it would be with a flat façade. The horizontal slats of western red cedar, like depilated eyebrows, have to accentuate the extensiveness of the five lingering apartment buildings. This all contrasts sharply with the façade on the northeast side which is built from a mixture of red brickwork. As a consequence, the project has two faces – a warm and a cool one, a dark and a light one, a vertical and a horizontal one – through which the oxymoron "ground-bound apartment" remains a contradiction in its appearance as well.

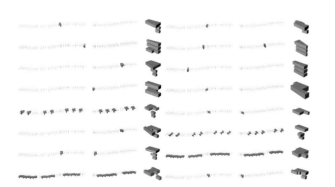

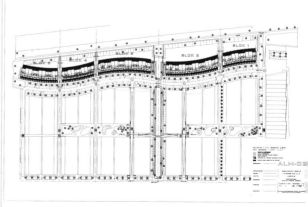

我们的每个住宅项目都试图以轻松的方式摆脱地产开发商最为看重的标准平面,创造出艺术品。这一点在Almere地区"运河之间"公寓项目中得到了很好的体现。当地的城市开发已经延伸到了"霍格运河"地区边绵延400米长的旧居住区内,该区域的住宅由119套公寓组成,公寓楼东北的人行道使楼群呈开放状态。由于整个街区的建筑只有三层高,我们建议每座公寓楼都在一层布置独立的入口大厅,以减少走道和电梯的数量,尽量降低服务成本。为了开放不同楼层的公寓,给客户提供多样化的选择,我们还沿进深15米的大厅为不同类型的公寓(共21种)设计了马赛克贴面。

所有公寓都在一层有独立入口,但这并不是该项目的唯一特别之处。令人惊讶的是,119个小型私人花园为每一个居民提供了高度私密化的空间:5米多长的木板靠近芦苇池;公寓立面没有出现额外的储藏室小门(这是一条旧的建筑法令,119个前门排成一排,足够人们出入);公寓里出现了温室和露台(以前有这样的公寓么?);某些"浮动"的房间漂浮在一层入口和二层房间之间;当然还有更特别的:前排的房子通常有三层,其中最大的一层面积达到186平方米。

当地的城市布局决定了该项目应该具有生态性的外观,例如,可以用木材覆盖临"格罗特运河"的立面,用立面部件形成一个开放的板条结构(但这样做怎么可能会有与众不同的效果呢?)。然而,我们想用乳白色玻璃幕墙在建筑类型上制造混乱并反射出芦苇和河水,这样做也符合生态要求。垂直的防护墙玻璃有着轻纱般的效果,这比平板化的立面好得多。水平向的西部红雪松木板就像被剃掉的眉毛,突出了五座绵延的公寓楼的巨大体量。这一切都与建筑北面的红砖立面形成鲜明的对比。这样一来,该项目就有了两个截然不同的立面:一个温暖、一个冷酷、一个暗淡、一个明亮,一个呈竖直状、一个呈水平状。整座建筑虽然"紧贴地面",但在外观上还是有诸多看点。

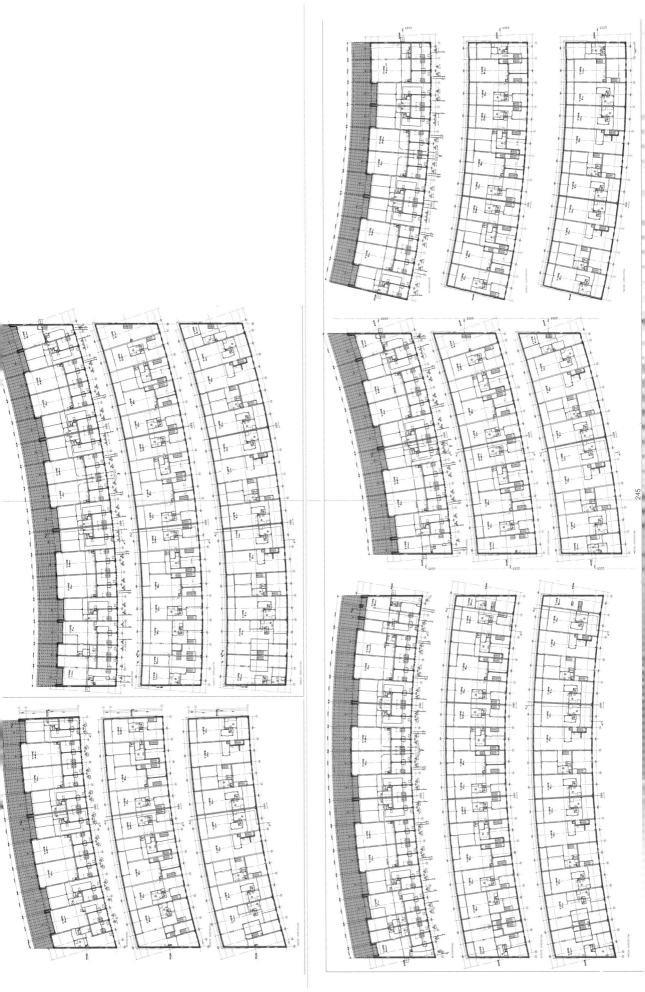

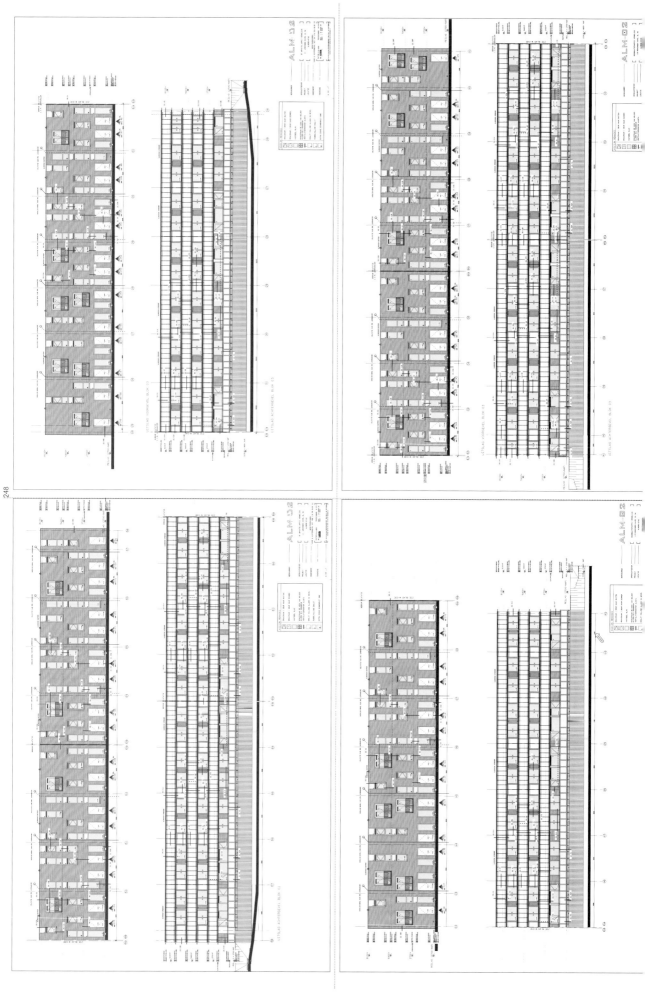

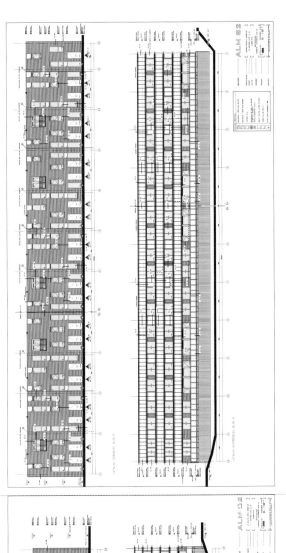

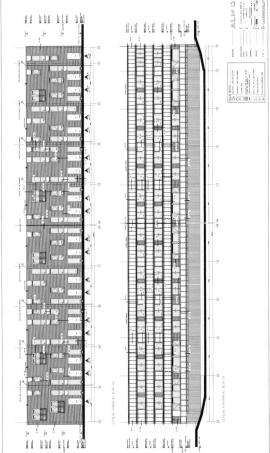

WWIIORMHOLE

Museum of the Second World War

Address: Zamczysko, Gdanks, Poland
Design: NIO Architecten
Structural engineer: Buro Happold, Warsaw
Client: Museum of 1939
Design team: Giacomo Garziano, Maurice Nio, Arek Seredyn
Start design: 2010

虫洞
二战博物馆

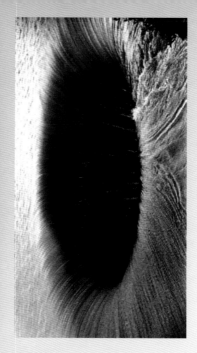

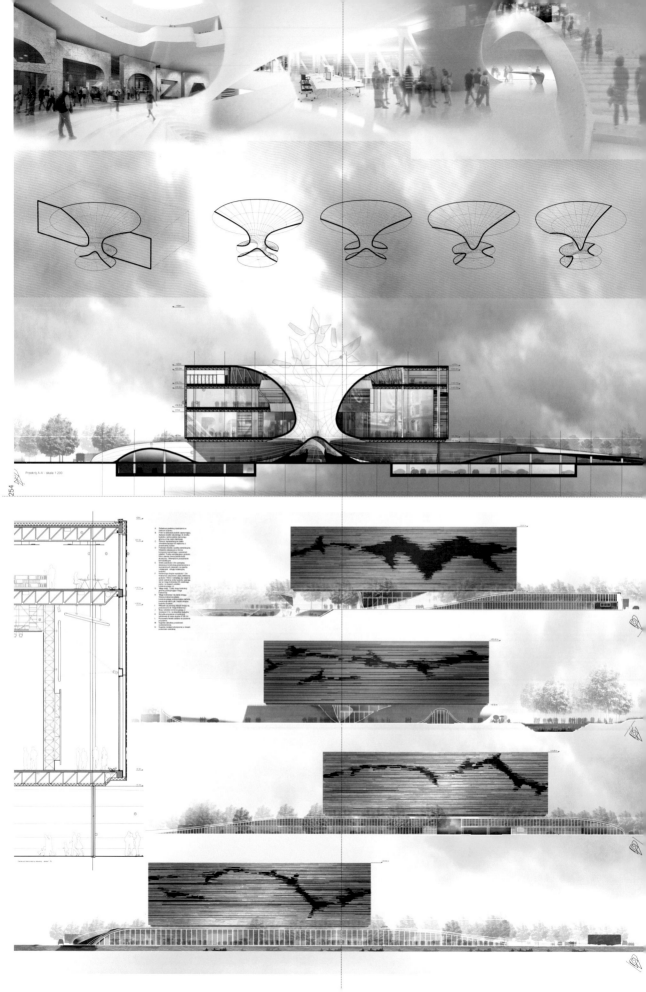

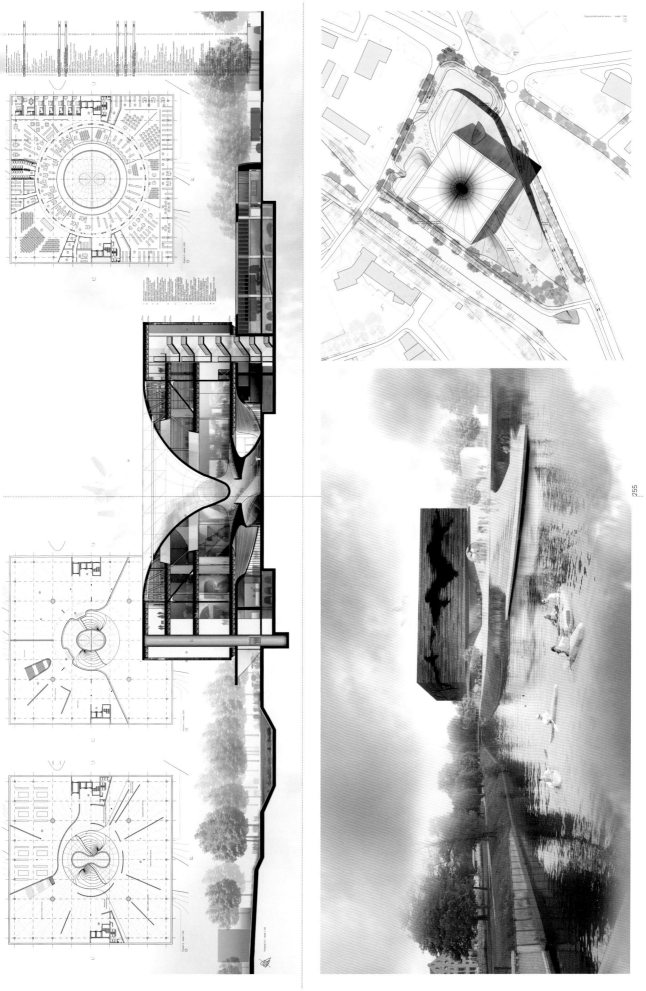

Waste Incinerator, Twente, the Netherlands

Bart Lootsma
(Originally published in Domus 791, March 1997)

垃圾焚烧场，荷兰特温特

Bart Lootsma

Bart Lootsma (Amsterdam, 1957) is a historian, critic and curator in the fields of architecture, design and the visual arts. He is a Professor for Architectural Theory at the Leopold-Franzens University in Innsbruck and Guest Professor for Architecture, European Urbanity and Globalization at the University of Luxemburg. Before, he was Head of Scientific Research at the ETH Zürich, Studio Basel, and he was a Visiting Professor at the Academy of Visual Arts in Vienna; at the Akademie der Bildenden Künste in Nürnberg; at the University of Applied Arts in Vienna and at the Berlage Institute in Rotterdam. He held numerous seminars and lectured at different academies for architecture and art in the Netherlands.

Bart Lootsma was guest curator of ArchiLab 2004 in Orléans and he was an editor of ao. Forum, de Architect, ARCHIS, ARCH+, l'Architecture d'Aujourd'hui, Daidalos, DOMUS and GAM. Bart Lootsma published numerous articles in magazines and books. Together with Dick Rijken he published the book ,Media and Architecture' (VPRO/Berlage Institute, 1998). His book 'SuperDutch', on contemprary architecture in the Netherlands, was published by Thames & Hudson, Princeton Architectural Press, DVA and SUN in the year 2000; 'ArchiLab 2004 The Naked City' by HYX in Orléans in 2004.

Bart Lootsma is Board Member of architektur und tirol in Innsbruck and reserve-member of the Council for Architectural Culture at the Cabinet of the Austrian Prime Minister in Vienna. was a member of several governemental, semi-governemmental and municipal committees in different countries, such as the Amenities Committee in Arnheim, the Rotterdam Arts Council, the Dutch Fund for Arts, Design and Architecture, Crown Member of the Dutch Culture Council, Member of the Expert Committee 11. International Architecture Biennale, Venice 2008, at the German Ministry for Building and Planning as well as curator of the Schneider Forberg Foundation in Munich.

Housing - "The Cyclops"

The countryside on the eastern borders of Holland near the town of Hengelo is truly strange. At first sight, it looks quite open and rural, with its meadows and, here and there, copses and small farmsteads. Suddenly, however, one sees peculiar-looking structures that remind one of a miniature version of the wooden drilling rigs you see in some cowboy films. On the Boeldershoek terrain, these buildings become more frequent, variations occur and they rapidly increase in size. There are sheds with a very odd shape, most of them made of corrugated iron, their industrial function is nothing if not mysterious. Everywhere, there are electricity pylons. Along the banks of the canal huge factories loom; their cooling water makes for a dense mist that hangs over the water all through the winter. The dominant activity in this whole district is salt extraction. The salt is diluted by the water in underground strata after which it is pumped to the surface, where it is condensed. The scale and intensity with which the salt is drained here imply that on occasions tracts of land suddenly cave in, with everything that stands on them. It is therefore strictly forbidden to build anything within a radius of 100 meters of a drilling rig. What is more, so much garbage has been dumped in this region during recent years that strange-looking flat hillocks have been created. Sometimes vegetation has grown over them once more, others are completely covered with black agricultural plastic; others again have been partially dug out, with blue pieces of plastic – apparently the toughest sort of plastic– placed everywhere in the pitch-black layers. Sometimes one even glimpses the silhouette of an enormous waste truck with its container tipped upwards. This countryside is no longer just a metaphor for how we humans have been behaving towards the planet during recent years; rather it is pars pro toto, the part that implies the whole. Perhaps this is the reason why we haven't yet found a name for it.

In recent years, people have increasingly come to realize that we can't go on like this. In most of Holland, for instance, for some years now organic and inorganic waste have been separated, with the former being made into a sort of compost. Starting in 1996 dumping waste has been prohibited in Holland. In the context of this prohibition seven or eight AVIs have been built on strategic sites in different parts of the country. AVI stands for Afval Verbrandings Installaties, or Waste Incineration Plants. Waste matter is collected and burnt in these installations; in the process as much of the residual substances as possible is either made productive once more or else they are recycled in the environment in purified form. The most recent addition to the series of AVIs is that in Twente. It is a gigantic machine into which the waste is poured and which produces slag and fly ash and enough electricity to supply a middle-sized provincial city. Since this activity is also a productive one, the plant is also intended to be completely self-supporting and even profitable. Half of the machine consists of an installation for purifying the gaseous smoke so that what comes out of the chimneys is in many respects cleaner than the air outside. This indicates just how much has changed in our society and just how much we are prepared to invest in our environment. Even so, these AVIs only represent an intermediate stage, one step towards a society that will produce much less waste, because most of it will be recycled. The expectation is that it will be twenty years before this is the case; this is also the write-off period in the planning of the AVIs.

Maurice Nio occupies an exceptional, not to say unusual, position in the world of Dutch architecture. On the one hand, he works for a large firm, Bureau De Gruyter; that has a reputation for being highly commercial; on the other, he works for the very small avant-garde firm of NOX. At present, he is also active with the firm of VHP traditionally concerned with the landscape and with city planning, where he is setting up a brand new architectural department. Besides that, he is a famous video artist, and at this year, his collected essays on architecture, film and art will be published, with the title You Have the Right to Remain Silent. All in all, he would seem to be someone who is difficult to categorize, the sort of person who can easily switch a marginal position for one that is right in the mainstream and vice versa.

Retail Park in Roermond

In actual fact however categorizing him in this way is not entirely accurate, because everything Nio does displays a high degree of consistency -everything links up with everything else. In this connection it is interesting to note that Maurice Nio is the Dutch translator of two books by Jean Baudrillard, namely Amérique and Les stratégies fatales. He would appear to be deeply affected by Baudrillard's aim of assailing the obscenity of the world we live in with its own means. "In opposition to that which is truer than true, we propose that which is falser than false", Baudrillard wrote in Les stratégies fatales. "We will not oppose the visible and the invisible, instead we will look for what is uglier than the ugly - namely the monstrous. We will not oppose the visible and the hidden, we will look for that which is more hidden than the hidden: the secret. We will not look for change and oppose what is motionless and what is mobile, we will look for what is more mobile than mobility, namely metamorphosis.... We will make no distinction between the true and the false, we will look for that which is falser than the false -namely illusion and appearances... In this increasingly extreme situation, we should separate the effects of obscenity and those of seduction in a manner that is possibly radical, but it may also be that we will have to bring them together, uniting them".

Waste incineration plant in Hengelo

This is the strategy that Nio has applied in the Twente AVI building. It is set in the middle of the weird post-industrial landscape of Boeldershoek and it is by far the largest building there. Its design fits in with that of mysterious closed sheds and structures in its surroundings, but it is much bigger. It swells and bulges with the installations that comprise it; its shape changed during the building process and it has only now solidified. It is the epitome of what Rem Koolhaas calls "Bigness"; it is a building that one can only begin to describe in terms of figures that are simply staggering: 230,000 tons of waste per year, development costs 600,000.000 guilders... Its relationship to the buildings around is that of a medieval cathedral to the city that surrounds it; it has every appearance of having been built in the same spirit, but it is larger and its detail is more complex. And moreover, the function of this building is in inverse proportion to that of the other buildings in its surrounding. From a distance, the Twente AVI seems to float free of ground as though it might slowly start to move forward, sucking up everything in its path. Nio calls it a gigantic dung beetle with machinelike tendencies, a sort of metallic green-coloured insect that feeds on garbage, being assisted in this by other coleoptera. Nio as referring here to the six smaller buildings that surround AVI itself which he also designed: the condenser for reclaiming water from steam, the slag building where the hard metals and mineral residue from the incineration process are pulverized, the gas reduction station, the weighhouse for weighing the trucks, the pumping station next to the pond, with its three gigantic pump for supplying the fire-extinguishing systems for the building and the office building. Apart from the installations themselves, Nio has designed everything on this terrain right down to the furnishings. He gave particularly detailed attention to the chimney. Normally, this would consist of some funnels sticking up into the air; here, however it is the pièce de résistance that sums up the essence of the whole project.

兰东部边境的亨厄洛镇有一种奇特的景观：随处可见的小型 场让那里乍看起来好像一片开放的乡村，然而，一些特殊的构 物却让人联想起西部电影中的小型的木质钻机。在Boelder- hoek地区，这种构筑物更为普遍，体量也更大一些。这种形状 特的棚子大多由瓦楞铁制成，除了制造神秘感之外并无它用。 当地，电缆塔随处可见，沿运河河堤布满了大型纺织厂，这些工 排放的冷却水使得冬季的河面笼罩在薄雾中。整个地区主要的 产活动是萃取盐。人们利用地下水将盐稀释后再将其泵到地面 行浓缩。盐的流失程度与当地突发性大规模土地塌陷有关。因 ，在钻机半径100米范围内严禁盖任何建筑物。随着近些年大 的垃圾在当地囤积，出现了很多垃圾山。有的小山上会长出植 ，有的完全被黑色农用塑料覆盖，还有的小山被再次挖开，漆黑 断面露出蓝色的塑料（最难分解的一种塑料）。有时我们甚至能 到上面有垃圾车的轮廓。这里的景观不仅隐喻了人类近些年来 地球的所作所为，它还暗示着整个pars pro toto。这也是我们 法为它命名的原因。

些年来，人们越来越清楚地认识到再这样下去是不行的。在荷 ，垃圾分类已经进行了很多年，而且有机废弃物被制成了肥料。 1996年起，倾倒垃圾在荷兰被禁止，全国不同地区的中心已建 七八个AVI。AVI是Afval Verbrandings Installaties（垃圾 烧场）的缩写。收集来的垃圾在AVI中进行焚化，剩下的残留 要么再次投入生产，要么以纯净物的形式回归自然。最近修建 一座位于特温特的AVI，垃圾被倒入巨大的机器中，焚烧过程 产生炉渣、粉煤灰和可供一个中等省级城市使用的电力资源。 一有益的创举使得垃圾焚烧场不仅可以完全自给自足，甚至还 可能盈利。焚烧场一半的设备是用来净化的气体烟雾的，经处 后排出的气体比户外的空气还要清洁。这些事实表明，我们的 会已经改变了很多，我们为改善环境也投入了很多。即便如此， 些AVI也只是垃圾处理的中间阶段，我们的社会将更进一步， 来的大部分垃圾将被回收再利用，废物总量会比现在少得多。 们期望在未来20年达到这一目标，到那时我们就不用规划建设 VI了。

aurice Nio在荷兰建筑界享有很高的声誉。一方面，他为Bu- au De Gruyter这样享有高度商业信誉的大公司工作；另一方 ，他为前卫的NOX公司工作。目前，他积极与VHP（擅长景观和 市规划）合作，在那里建立一个全新的建筑设计部门。此外，他 是著名的影像艺术家。他的新书《你有权保持沉默》收录了他 建筑、电影和艺术的观点和文章。总而言之，我们似乎很难给 aurice Nio定位，他的地位介于主流和边缘之间。

实上，这种分类不完全准确，因为Nio的所有作品都具有高度的 致性，它们紧密地联系在一起。我们应该关注Maurice Nio翻 的由Jean Baudrillard撰写的两本书——《美》和《致命的策 》。他似乎十分赞成Baudrillard的看法，用丑恶所特有的方式 攻击丑恶本身。Baudrillard在《致命的策略》中写道："我反 更为真实的事实，我赞成更为虚伪的假象……我不想寻求可见的 物和隐藏的事物对立起来，我想寻找最丑恶的事物、最为隐秘 事物，寻找秘密。我不想寻求变化，将移动和静止对立起来，我 寻找比流动性更善变的变化……我们不想区别真与假，而要寻 更为虚假的错觉和表象……在极端情况下，我们应该把丑恶和 感的影响尽可能分开，但有时也必须把它们放在一起，结合起 。"

就是Nio在特温特AVI设计中应用的策略。这座建筑位于后工 化的Boeldershoek景观中，是那里迄今为止最大的建筑。它 设计和结构与当地神秘而封闭的棚状建筑相适应，但体量要 得多。大型设备使建筑体量膨胀起来，整个建筑的外形在建设 程中一直在改变，直到现在才定型。它是Rem Koolhaas所说 "大"的缩影，描述该建筑的数字煞是惊人：每年消耗23万吨 ，开发成本高达60万荷兰盾……它与周围建筑的关系像是一个 世纪大教堂与周边城市的关系，AVI在外观设计上和大教堂一 具有统一的构思，但其体量更为巨大，细节也更为复杂。而且这 建筑的功能与周围的其他建筑物完全相反。从远处看去，特温 AVI似乎可以自由地浮动，就好像它能慢慢行走，吸入路上的 切东西一样。Nio称该建筑像一个巨大的具有机械般外表的甲 虫，一以垃圾为食的绿色金属昆虫。这只昆虫在处理垃圾的 程中，还得到了其他甲壳虫的协助。Nio在这里所说的其他甲壳 ，是环绕在AVI周围的六座小型建筑物（也由Nio设计）：回收 凝水的冷凝器、粉碎焚烧过程中产生的金属和炉渣的熔渣楼、

气体吸收站、货车过磅处、池塘边的水泵房及其附带的三个大水泵（为AVI和办公楼提供消防用水）。除AVI之外，Nio还设计了基地中的所用东西（包括室内家具），并特别精心设计了烟囱：一般情况下，烟囱会直立伸向空中，而该项目中的pièce de résis- tance则表现出整个项目的精髓。

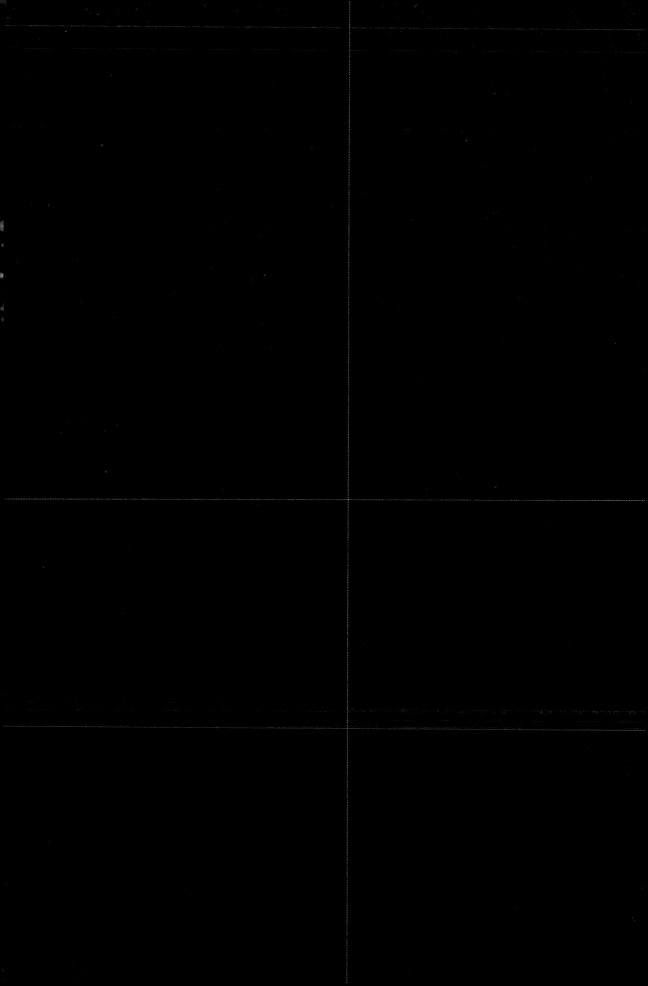

Unspeakable
Shaping Things that Cannot Be Told

无法言说
无法用语言表达的创建

Madame Butterfly
Dark Matter
Eat This
The Fire Emperor
Snake Space
Wait until Dark
Watermark

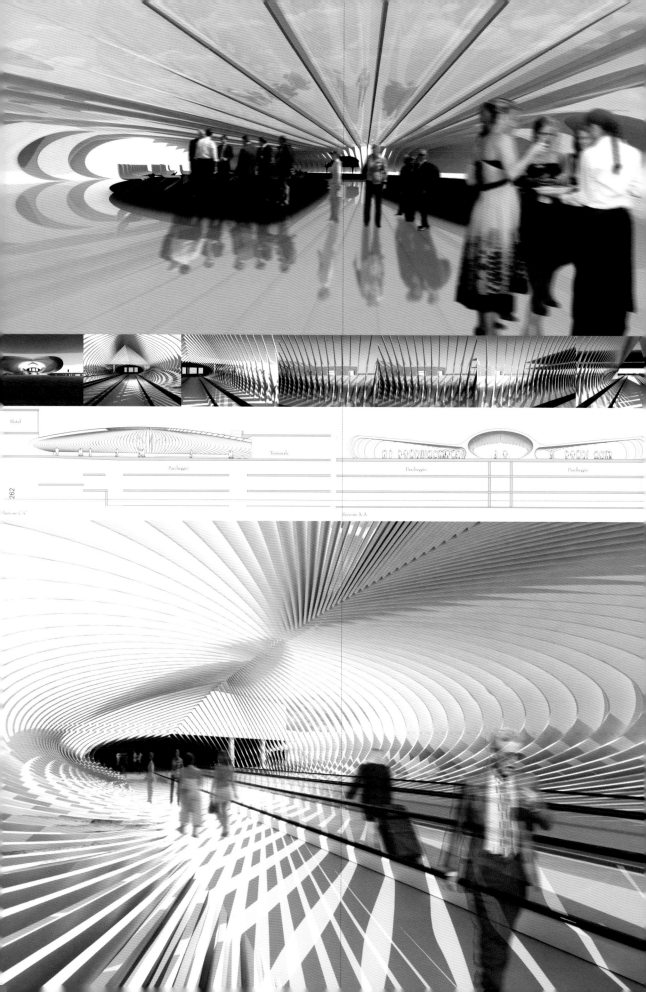

MADAME BUTTERFLY

Building Entrance for Terminal 1 of Malpensa Airport

Address: Terminal 1, Airport Malpensa, Italy
Design: NIO Architecten
Contractor: Consorzio Etruria
Design team: Joan Almekinders, Alexander Hertel, Maurice Nio
Start design: 2009
Costs: euro 2,500,000

蝴蝶夫人

米兰Malpensa机场1号航站楼

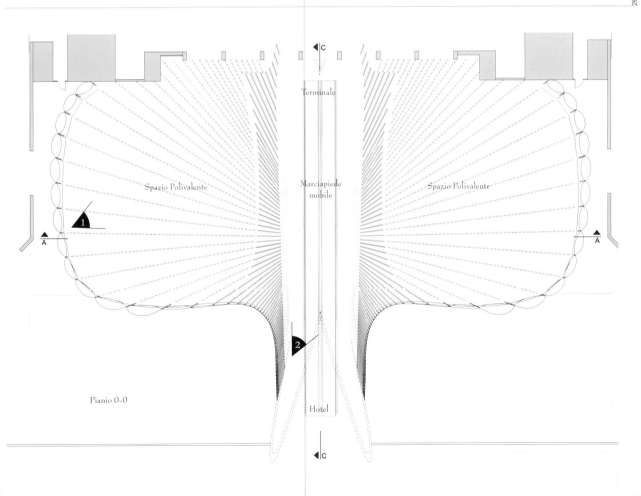

Pianio 0-0

DARK MATTER

Traveling Installation

Address first location: Temporary Museum Pecci Milano, Italy
Design: NIO Architecten
Design team: Joan Almekinders, Giacomo Garziano, Maurice Nio, Luca Rimatori, Jan-Willem Terlouw
Start design: June 2009
Completion: April 2010

Photo credit: Milano Adelaide Corbetta, Milano Wais Wardak

暗物质

巡回装置

We live in a world that is becoming more and more transparent. Riddles are being solved, secrets uncovered, irregularities glossed over, twists rationalized: the other becomes the same. At the point of complete transparency, we end up in a state of complete obscenity as well. After all, when all veils are gone, we are left with nothing but trivial nudity. The temptation has disappeared.

At the same time, however, there is the awareness that 90% of our universe consists of dark matter. Matter that could well be the cause of many incomprehensible phenomena in the universe. If you apply this hypothesis to our world, that would mean that fortunately only 10% could really be transparent and obscene. So let us search for the rest, for that 90% dark matter.

Dreams, emotions, plants, stones, but especially animals are part of that 90% dark matter. Animals do not talk. They growl, hiss, whistle, bark. We can hear them, but not understand them. That in itself makes them a mystery. What is more beautiful than staring at an animal that does not say anything? What is more beautiful than an animal that does not communicate and withdraws into its own shadow?

(The work of art is a 17-meter-long polyester object of 8 animals that have melted together and that are hung with about 15 screens on which images of animals can be seen, recorded by webcams mostly at exotic locations.)

我们生活的世界正变得越来越透明化。人们发现秘密，揭开谜底，为不合理的东西寻找理由：其他的东西也是如此。当一切都完全透明的时候，我们也将处于一种猥琐的状态中。毕竟，当所有的面纱都消失的时候，诱惑就消失了，只剩下没有价值的躯体。

然而，宇宙的90%都是由暗物质组成的，这种物质可能是许多谜团的原因所在。根据这一假说，我们的世界就只有10%的物质是真正透明和肮脏的。让我们来寻找那90%的暗物质吧。

梦想、情感、植物、石头，特别是动物，有90%是由暗物质组成的。很多动物虽然不说话，但它们会发出咆哮声、嘶嘶声、口哨声和犬吠声。我们可以听到它们的声音，但不明白其中的含义。这使得它们本身就成为了谜团。注视一只不说话的动物是这世上最美丽的举动。一只不说话、蜷缩在自己的影子中的动物是这世上最美丽的事物。

（该艺术品是一个17米长的聚酯纤维动物雕塑，有8只动物连接在一起，整个雕塑与15幅动物图片一起悬挂起来，大多数图片是在国外拍摄的。）

EAT THIS!

Boiling point of the public domain

Editors: Joan Almekinders, Maurice Nio
Publisher: 1001 Publishers, Amsterdam
Text: Philip Mechanicus, Maurice Nio, Joan Almekinders, a.o.
Photography: Otto Snoek
Design team: Joan Almekinders, Georg Bohle, Stefano Milani, Maurice Nio,
Alexander Paschaloudis, Arek Seredyn
Grafic design: Derk Reneman & Manulea Porceddu
Press: De Maasstad, Rotterdam
Start design: 2005
Publishing: 2006

吃掉它!

公共空间的沸点

EAT THIS! is a call to build market halls in the Netherlands that enrich the public domain.

The authors (NIO Architects) designed "an inimitable gorge machine, a hedonistic jumble of spaces (...) with interiors from all kinds of nationalities and styles": the Fire Emperor. The Fire Emperor, a new appearance in the heart of every major Dutch city, the new non-stop market hall, a pivot of public life.

They have also cooked up 10 new food formulas and highlight 25 famous market halls in world cities.

Philip Mechanicus, photographer and culinary journalist, serves up a number of food themes in his own unique way.

22 photographs by photographer Otto Snoek have been added, snap shots of people (mainly) eating.

The book reads and looks like a market hall, a colourful collection of everything to do with food – the products, the preparation, the consumption. Wandering through the book, you will encounter the most weird and wonderful things, just as you would if you were strolling around a large market.

"Food makes people come out of their shell, only food is capable of delivering such delicious public experiences."

"吃掉它！"是荷兰的一个市场大厅，它的出现丰富了当地的公共空间。

Nio事务所的建筑师设计了"一个独一无二的重要机器，一个充满享乐主义和各国风情的内部空间"：热情的皇帝。 热情的皇帝"是出现在荷兰各主要城市的一张新面孔，这是一个永不停止运转的市场大厅，也是公共生活的中心。

人们还公布了10个新的食品配方，突出了25个世界大城市的著名市场。

Philip Mechanicus，作为摄影师和美食记者，以自己独特的方式撰写有关食物的文章。

文中插入的22张照片是由摄影师Otto Snoek拍摄的，主要是人们吃饭时的快照。

读起本书，就像进入了市场大厅，那里有丰富多彩的事物以及一切与食品有关的东西——原材料、烹饪和消费。阅读本书，你将遇到很多怪异而奇妙的事情，就像你亲自在一个大市场里闲逛一样。

"食品使人们走出家门，只有食物能够传递如此美味的公共经验。"

THE FIRE EMPEROR

New market hall, Rotterdam

Address: Grote Markt, Rotterdam, Holland
Design: NIO Architecten
Design team: Joan Almekinders, Georg Bohle, Stefano Milani, Maurice Nio, Arek Seredyn
Start design: 2004

热情的皇帝

鹿特丹新市场大厅

We want Holland to become acquainted with the Fire Emperor, a new holding of markets that has taken the shape of a building. The Fire Emperor will be the pivot of public life. No-one can escape from this voracious building that day and night devours anything that comes near: innocent tourists and experienced gluttons, pale potatoes and fresh coriander, tame pigeons and live squid, raunchy market stalls and exclusive restaurants, worn-out musicians and erotic services, the rotten smell of durians and the faded perfumes of waitresses. Everything is being digested, pushed along and shovelled out again, but not before it has been substantially reshaped under high pressure. Because it is the new non-stop market hall that houses in the Fire Emperor. Everybody has something to look for here, and everybody and everything is prepared to defy the heat of the kitchen in the Fire Emperor. This is where the city boils and where people travel along on the vapours of exotic dishes.

The Fire Emperor has no transparent airy hall on the inside, but is one big, inimitable devouring machine, a hedonic tangle of spaces inspired on intestines. The dazzling diversity of the food on offer is reflected in the festive spirit of the interior design. It is as if the market hall itself has been fattened with interiors that have all nationalities and styles. A corny German-expressionist space is being alternated with an obscure Thai-New Age room. It is a market hall that is not embarrassed of anything and that burps nicely after having consumed yet another stylistic tour de force.

Every big city, both inside and outside the Netherlands, deserves a Fire Emperor. You could also say that every big outside market should go together with a Fire Emperor, because the "covered market" and the "outside market" are not opposites fighting for the same bone, but they are a couple that feeds themselves upon flows of food and consumers. What is more, because they are complementary they make their surroundings brew again. In Rotterdam, for example, we have proposed to locate the Fire Emperor on "het Steiger", a spot that, before the war, was one of Rotterdam's most mysterious places but now looks desolate and deluded. We have placed the Fire Emperor in such a way that it becomes the pivot between the existing outside market on "de Binnenrotte" and the new floating market on the water of "het Steiger", and this way, it lets things steam and damp nicely, day and night.

每一个大城市（不管是否在荷兰境内）都应该拥有"热情的皇帝"。每一个大型的露天市场都应该与"热情的皇帝"结合在一起，因为"室内市场"和"露天市场"并不是对立的，它们可相互补充食物和客流。更重要的是，它们的互补会使周边地区再次繁荣起来。例如，在鹿特丹，我们想在"het Steiger"地区建造热情的皇帝，该地点在战前是鹿特丹最神秘的地方之一，但现在却一片荒凉。我们在此修建热情的皇帝，使它成为现有的露天市场"Binnenrotte"与当地新的水上市场之间的支点。这样一来，当地的饮食业会迅速发展起来。

我们希望将"热情的皇帝"这种新型市场引入荷兰，它将成为公共生活的中心。没有人能逃离这个贪婪的建筑，它夜以继日地吞噬着靠近它的任何东西：无辜的游客、经验丰富的食客、灰白色的土豆和新鲜的香菜、驯服的鸽子和鲜活的鱿鱼、肮脏的大排档和独立餐馆、穿着破旧音乐家和色情服务女郎、烂臭的榴莲和褪色的女服务员香水。一切事物在高压下都会变质，但在那之前都要被一次次的消耗、转化和逐出。热情的皇帝中，每座小建筑都永不停止运转。每个人都能在这里找到所需要的东西，每个人、每件东西都准备好抵抗热情的皇帝内部高温烹饪产生的热气。这里是城市的沸点和人们呼吸异国菜香味的地方。

"热情的皇帝"内部没有通透的大厅，它是一个独特的吞噬机器、一个享乐主义的空间。充满节日气氛的室内设计映衬出琳琅满目的食物，市场大厅多种民族风格的内部装饰本身就非常丰富。一个德国传统表现主义的空间被不起眼的泰国新时代风格替代了。这个市场不会为任何事感到难堪，而且乐于让顾客感受到不同国家的不同风格。

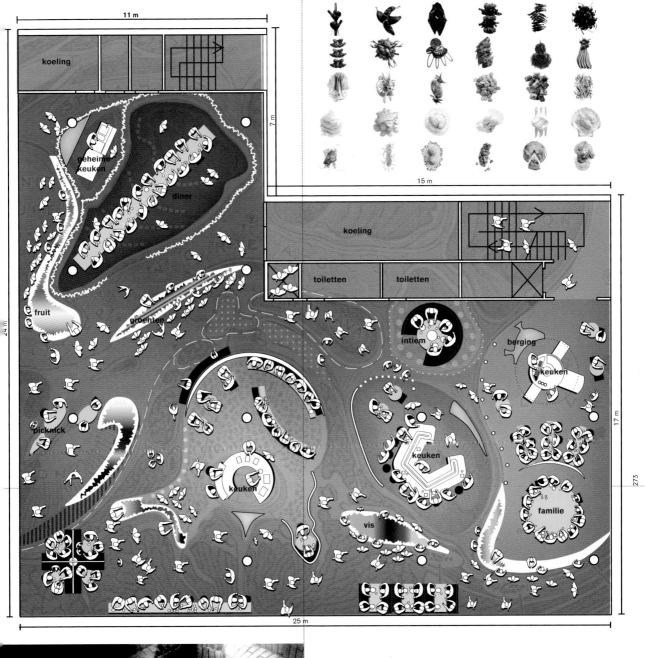

From the opposite perspective, we think, in line of Rupert Sheldrake's ideas, that the human spirit is not confined to the head, but part of a much bigger spirited reality. The spirit is a virtual body that can leave our biological body and get in touch with other virtual bodies, spirits, forms of consciousness and, as a consequence, also with animals and natural spirits, angels and demons, pokémons and fantasies. This ability of the spirit to be able to wander outside the body has literally been minimized by Descartes in the 17th century. Ever since, the sphere of influence of the spirit has been limited to the brains and the body is seen as a spiritless machine. Ever since, the body has become a technical space.

Precisely, on the intersection of the present technical space of our body and the technical space of our urban landscape, there is a possible new space. Snake Space. Where everything is soulless, where everything has become technical and/or functional to the bone – both body and space – it is possible through careful architectural interventions to give the relationship between persons and objects a spiritual dimension again. It is as if, in the space, other spaces are created, spaces that are not just made for people, but also and especially for spirits, for the traffic of the virtual bodies. They are the lift shafts of our thoughts, the ventilation channels of our projections and fantasies, and they are, of course, also the attics and cellars of our consciousness.

从相反的角度来看，我们赞成Rupert Sheldrake的观点，认为人文精神并不是单纯的思想，而是属于一个更大的精神实体。这种精神可以超越我们的躯体与其他机体、精神、意识，乃至动物和自然的精神、天使与恶魔、精灵与幻想相接触。精神实体的这种能力，在17世纪被笛卡尔扼杀得最为严重，其影响范围被限制在了大脑中，身体则变成了一个没有灵魂的机器，被看成是一个技术空间。

恰恰在身体和城市景观的交点上可能会出现一种新的空间——蛇形空间。在人的身体和城市空间内部，所有的东西都没有灵魂，从本质上都属于技术或（和）功能的范畴。我们可以通过建筑的干预，为人和事物之间的关系再次赋予一层精神方面的尺度。比如我们创造的空间，不仅为人服务，也为精神实体、人的交通需求而服务。这些空间可以是装载我们思想的升降机、充满人类预测和幻想的通风道，当然也可以是充满我们意识的阁楼和地下室。

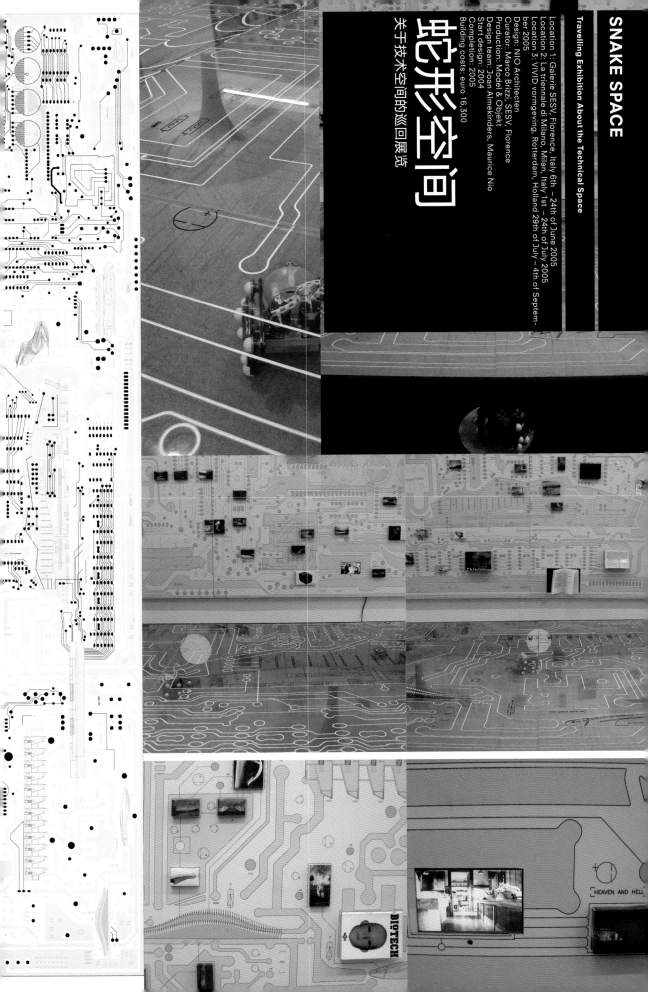

SNAKE SPACE

Travelling Exhibition About the Technical Space

蛇形空间
关于技术空间的巡回展览

Location 1: Galerie SESV, Florence, Italy 6th – 24th of June 2005
Location 2: La triennale di Milano, Milan, Italy 1st – 24th of July 2005
Location 3: VIVID vormgeving, Rotterdam, Holland 29th of July – 4th of September 2005
Design: NIO Architecten
Curator: Marco Brizzi, SESV, Florence
Production: Model & Objekt
Design team: Joan Almekinders, Maurice Nio
Start design: 2004
Completion: 2005
Building costs: euro 16,300

WAIT UNTIL DARK

Lighting Plan for Tallinn, Estland

Address: Tallinn, Estonia
Design: NIO Architecten
Client: NODI
Design team: Joan Almekinders, Stefano Milani, Arek Seredyn, Alexander Paschaloudis
In partnership with: Raul Kalvo, Kadri Kerge, Juhan Rohtla, Tõnis Savi, Johanna-Mai Vihalem
Start design: workshop 18–21 november 2004

等到天黑

Estland塔林灯光设计

5-HT
serotonin 400mg
gives a clear, alert outlook
with a higher awareness function

Get a

Nion

light

WATERMARK

Exhibition for 'Perimeters, Boundaries and Borders'

Address: City Lab, Lancaster, England
Design: NIO Architecten
Client: Folly, Fast-UK, Arts Council England
Manufacturer: Gravotech b.v. Specialist in 3D
www.Gravotech.nl
Design team: Joan Almekinders, Radek Brunecky, Sean Matsumoto, Maurice Nio
Start design: July 2006
Completion: September 2006

水印
边缘、分界线和边界的展览

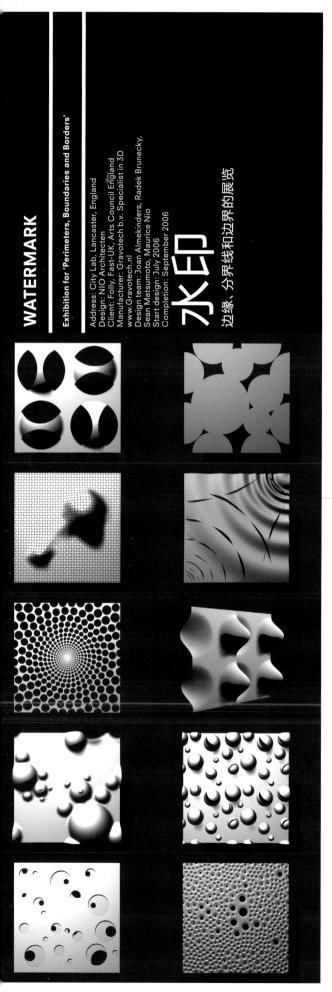

Let's face it; things are never what they seem to be. There is no such thing as "dead" or "blind" matter. The fact that we, people, cannot perceive any sign of consciousness in – for example – a stone, doesn't give us the right to say that there is no consciousness in stones, just because we cannot recognize it. The emotional bond between people and stones, between subject and object, is the base for every experience. Something makes you moody, another thing makes you happy and nothing is unbiased. We are affectionate creatures and we experience things like the inhalation of oxygen. The only question is: Where is the emotion? Is it here or is it there? Is it intrinsic or extrinsic?

The 10 plates that we are presenting are mock-ups of façade panels for a cluster of buildings with divergent functions; a music hall, a national soccer museum, a fast food restaurant, a school, a wellness center and several outdoor activity shops, all located in Middelburg, a Dutch city close to the sea. The panels are embodiments of 10 moods that relate to leisure, crossed with 10 physical appearances that relate to water. Desire-whirl, arousal-cohesion, thrill-humidity, satisfaction-drop, curiosity-drifting, relaxation-rain, joy-floating, excitement-boiling, welcoming-wave, anticipation-ripple, these facades will become the mindset of the project, cut and pressed brutall in steel and aluminium.

Although the production of moods is fully artificial, every matter is capable to withdraw something so natural and yet so inconceivable as "emotion". Concrete, steel, stones and panels are like actors in a movie, trained to evoke emotions in a synthetic manner. We can direct this scene by loading a building full of sensation and packing a façade full of touch. The emotion can be simmering on the background, or bursting on the foreground, but it is always part of the experience. Our intention is to go beyond the senses of seeing and touching. It is a "built-in" temper. Or as the dea and blind Helen Keller wrote: "The best and most beautiful things in the world cannot be seen or even touched – they must be felt with the heart."

我们要面对这样一个现实：事物的本质远非它的表象。根本就不存在"死"或"失明"，只是我们无法理解某些意识的迹象罢了。例如，我们不能说一块石头没有意识，只是因为我们无法察觉到它的意识。人和石头之间、主体与客体之间的情感联系，是一切体验的基础。某些东西令你喜怒无常，有的东西令你快乐，所有的事物都是有感情色彩的。我们喜爱生灵，经历其他事物，这些经历与我们呼吸空气是一样的。唯一的问题是：情感在哪里？在这里还是在那里？在事物内部还是外部？

我们所陈列的10块面板取自10座不同建筑物的外墙：一座音乐厅、国家足球博物馆、一个快餐店、一所学校、一个健身中心和一些户外活动商店，这些建筑都位于荷兰沿海城市Middelburg。这10块面板通过不同的物理特性体现出10种情绪：旋涡代表欲望，凝聚力代表觉醒，湿度代表恐惧，滴落的水滴代表满意度，漂流代表好奇心，降雨代表放松，漂流代表喜悦，沸腾代表兴奋，波动代表欢迎，涟漪代表期待。这些立面意向将成为该项目的思维模式，我们通过切压钢铁和铝板将其变为现实。

虽然我们用人工的方式生产情绪，且每种情绪都可以与自然相联系，但"情感"仍然是一种不可思议的东西。混凝土、钢筋、石材和面板就像影片中的演员，我们以人工合成的方式唤起他们的情绪，通过对建筑立面的感受和触觉指导剧中的场景，将情感安置于背景中或让它在前景中迸发出来。但这种情感毕竟是体验的一部分，我们的最终目标是要超越感官的体验和触觉，将情感"内置"于性情之中。正如聋哑盲人作家海伦·凯勒所说的那样："世界上最好、最美丽的东西不是用视觉和触觉感知的——我们必须用心灵去体会它们。"

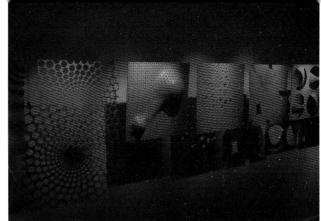

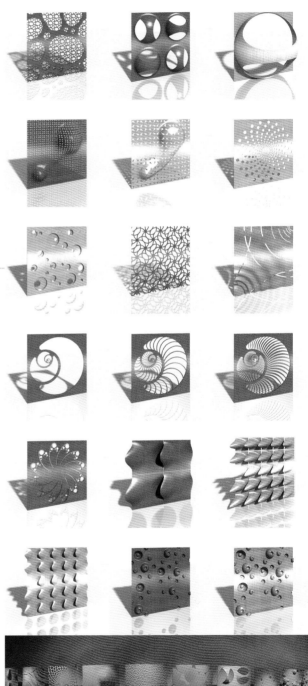

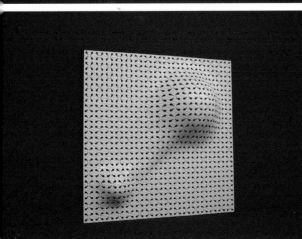

Maurice Nio
Joan Almekinders
建筑师简介

Curriculum Vitae

Maurice Nio

Joan Almekinders

Maurice Nio (1959) graduated cum laude as an architect in 1988 at the Faculty of Architecture of the Delft University of Technology on a villa for Michael Jackson, the most curious final project of that year. This project has been of vital importance to his hybrid approach. Through a mixture of mythological and pragmatic mental processes, cryptic and at the same time utterly transparent design strategies, he has realized projects at BDG Architekten Ingenieurs (1991–1996), such as the enormous waste incinerator aviTwente. At VHP stedebouwkundigen + architekten + landschapsarchitekten (1997–1999) he realized the Zuidtangent, the longest high-quality public transport line in Europe. As from January 1st, 2000, he operates from his own design studio NIO Architecten and currently works on the most beautiful shopping centre in the world and the most obscure houseboat in the Netherlands. Maurice Nio gives many lectures home and abroad, but has decided he no longer wants to be a teacher on art schools, academies or universities, and would like to design books again, or make his eighth video production, or else, write new articles on urban development, architecture, film, video, television, photography or dance. Achievements anyway are his books - You Have the Right to Remain Silent (1998) and Unseen I Slipped Away (2004).

Joan Almekinders graduated in 1994 with an honourable mention at the Faculty of Architecture at the Technical University in Delft, Holland. He collaborated as a freelance architect with OKRA Landschapsarchitecten, Lafour & Wijk, BRO and Robin Anderson, Miami. From June 1995 till April 2000, he worked at NOX Architekten at which the Freshwater pavilion at the island Neeltje Jans and the V2_lab in Rotterdam where two of his major projects. From April 2000, he was as an architect at the office of Hoogstad architecten involved in several projects such as the masterplan for the Technical University Twente and the preliminary design for the underground central railway station in Delft. Since June 2002, he works as an architect at NIO architecten. In January 2004, he became partner with the founder of this office Maurice Nio. Together with him, he directs an office of about 8 people, all architects, working on a wide range of projects such as commercial retail and leisure, infrastructural works, housing and cultural buildings. Next to this practice, he has been teaching since 1994 at the IAH Larenstein, Velp and from 2001 till 2004, at the European Masters for Landscape Architects, Velp-Wageningen. He has lectured in the Netherlands, the United Kingdom, Estonia and the United States.

Maurice Nio生于1959年，1988年以优异成绩毕业于代尔夫特理工大学，成为一名建筑师。他的毕业设计是为迈克尔·杰克逊设计的别墅，这是当年的最佳作品，也是他用混合设计法设计的建筑代表作。Maurice Nio融合了神话思维和现实思维，应用神秘却完全透明的设计策略，完成了一系列建筑作品，包括他在BDG Architekten Ingenieurs（1991-1996年）期间设计的巨型垃圾焚烧炉aviTwente，以及在VHP stedebouwkundigen+architekten+landschapsarchitekten（1997-1999年）期间完成的Zuidtangent（欧洲最长的高品质公共交通线路）项目。自2000年1月1日起，Maurice Nio开始经营自己的设计工作室NIO Architecten，目前他正在设计世界上最美的购物中心和荷兰最不起眼的游艇。Maurice Nio在国内外许多地方讲学，但也不再想在艺术院校或大学任教，而是希望重新出书、制作他的第八张录影光盘，或在城市开发、建筑、电影、视频、电视传媒、摄影或舞蹈方面写写文章。他的成就体现在《你有权保持沉默》(1998)和《我悄悄地溜走了》(2004)两本书中。

Joan Almekinders于1994年毕业于代尔夫特理工大学建筑学院，并获得了殊荣。他作为一个自由职业建筑师与OKRA Landschapsarchitecten、Lafour & Wijk、BRO and Robin Anderson、Miami等多家建筑事务所合作。自1995年6月至2000年4月，他在NOX Architekten工作，期间设计了Neeltje Jans岛上的淡水展览馆和鹿特丹的V2_lab。从2000年4月起，他在Hoogstad Architecten任建筑师，参与了特温特理工大学的总体规划和代尔夫特中央火车站地下部分的初步设计。自2002年6月起，Joan Almekinders加入了NIO Architecten事务所。2004年1月，他成为了事务所创始人Maurice Nio的合伙人。Joan Almekinders的工作室共有8人，项目涉及商业零售建筑、休闲设施、基建工程、住房和文化项目等广泛的领域。除建筑实践之外，Joan Almekinders自1994年以来一直在IAH Larenstein和Velp任教。自2001年至2004年，他在Velp-Wageningen教授欧洲景观建筑。不仅如此，他在荷兰、英国、爱沙尼亚和美国都进行过学术讲座。

2000–2009
Designs and projects for NIO architecten, Rotterdam

- Railway station Vesuvio Est, Striano, Italy, closed competition, 2009
- Redevelopment of Lungomare di Levante-Lotto Nord, Siracusa, Italy, closed competition, 2009
- Railway station Cessange, Luxembourg, Luxembourg, closed competition, 2008–2009
- Hairstudio Explicit Hair Design, Arnhem, sketch design, 2008–2009
- Mobility Park, Roermond, sketch design, 2008–2009
- Retail Park, Almere, sketch design, 2008–2009
- Spinoza Center, Amsterdam, study, 2008–2009
- Outdoor Center, Roermond, under construction, 2008–2009
- Childcare Center, Prato, Italy, competition, 2008
- 3 Bridges for pedestrians and cyclists in De Oude Dokken, Gent, Belgium, closed competition, 2008
- Pergola construction over the N201, Haarlemmermeer, under construction, 2008–2009
- Multipurpose building in the Waldorpstraat, Den Haag, study, 2008
- House for the family Hunting, Waalre, sketch design, 2007
- Underpass for pedestrians and cyclists, Delft, closed competition, 2007
- Railway Hanzelijn Oude Land, under construction, 2007–2008
- 'ReBlackpool', regeneration of the seafront of Blackpool, closed competition, 2006–2007
- Movable bridge Vrouwenakker, Liemen, project realized, 2006–2008
- Extension Centro per l'Arte Contemporanea Luigi Pecci, Prato, Italy, under construction, 2006–2009
- Redevelopment Shopping Center Lewenborg, Groningen, closed competition, 2006
- Freeway A4 - section Burgerveen-Leiden, under construction, 2006–2009
- Bridge Broekland - de Watertuinen, 's-Hertogenbosch, closed competition, 2006
- Lightweight covering of freeways, study, 2005–2006
- Interior fittings for Saton Optiek, Katwijk, closed competition, 2005
- Advertising pillars (30 metres high), Roermond, closed competition, 2005
- Houseboat, Amsterdam, preliminary design, 2005
- 'De Friesebrug', movable bridge and tunnel, Alkmaar, closed competition, 2005
- Retail Park, Roermond, project realized, 2004–2008
- Biomass power station, Hengelo, final design, 2004–2005
- 22 Bridges in the 'Watertuinen', 's-Hertogenbosch, project realized, 2004–2007
- 350 Houses in Geuzenveld, Amsterdam, study, 2004
- ZEP Leisure Park in Middelburg, under construction, 2004–2009
- Penthousing Schiedamse Vest 91-95, Rotterdam, study, 2004
- Gateway Art Project, Queens Drive Flyover/Rocket Junction, Liverpool, closed competition, 2004
- Service building for Micro Beton, Rotterdam, preliminary design, 2004–2005
- Hotel-restaurant, Harderwijk, study, 2004
- Interior fittings for 'The Village', Phase 3, life-style department store, Voorburg, project realized, 2003–2004
- House and pavilion, Waverveen, project realized, 2003–2005
- New design for the Europanels and Billboards of Viacom, study, 2003
- 'The New Hunting Grounds', proposal for a strategic intervention of food consumption in the public-domain, study, 2003–2004
- 48 Houses on the PontMeyer site, Zaanstad, contract drawings, 2003–2007
- Palace for the Queens of the Night, Rotterdam, competition, 2003
- 'Amsterdam 2.0', Phase 2, a vision of the future of the City of Amsterdam, study, 2003–2004
- Interior fittings for 'The Village', Phase 2, life-style department store, Voorburg, project realized, 2002–2003
- 5 Houses for screw jacks in Leidsche Rijn, Utrecht, closed competition, 2002–2003
- Tunnel, Pijnacker, project realized, 2002–2004
- Canopy NS station Hoofddorp, Hoofddorp, closed competition, 2002
- 33 Houses in Groenoord-Zuid, Schiedam, project realized, 2001–2005
- Bus station, Hoofddorp, project realized, 2001–2002
- Tunnel, Amstelveen, project realized, 2001–2004
- Showroom for Schaffenburg, Zwijndrecht, project realized, 2001
- 65 Detached houses, Leidschenveen, preliminary design, 2001–2004
- Two tunnels in Vathorst, Amersfoort, project realized, 2001–2005
- 20 Houses, Leidschenveen, contract drawings, 2000–200
- Office building, Utrecht, study, 2000
- Office building, IJsselstein, study, 2000
- Service building for MC Automatisering, IJsselstein, study, 2000
- Interior fittings for 'The Village', Phase 1, life-style department store, Voorburg, project realized, 2000–2002
- Redevelopment of the Runshopping Centre, Hoorn, study, 2000
- Sound barrier houses, Naarden, study, 2000
- Office building in Veenendaal, study, 2000
- Extension of the office building for Twence Waste Incineration, Hengelo, project realized, 2000–2002
- 198 Houses, Almere-Stad, project realized, 1999–2003
- Disembarkation installation for cruise ships, Amsterdam, preliminary design, 1999–2000

1997–1999
Designs and projects for VHP Urban Developers + Architects + Landscape Architects, Rotterdam

- Public transport junction and bus depot in Hoofddorp, study, 1999–2001
- Joinery works, businesses centre and offices in Ede, project realized, 1999–2001
- 'Expo 2001', 50 deregulated houses in Almere-Buiten, closed competition, 1999
- 'Batavia City', Factory Outlet Centre in Lelystad, project realized, 1999–2001
- Sound protection shields along the N22 and N201 roads in Hoofddorp, project realized, 1998–2002
- Viaducts and tunnels for the Central Section of the South Tangent in Haarlemmermeer, project realized, 1998–2002
- 'The Village', life-style department store in Voorburg, project realized, 1998–2001
- Auxiliary boiler house for the PNEM in Breda, closed competition, 1998
- The Central Section of the South Tangent, 14 public transport stops between Haarlem and Schiphol, project realized, 1998–2001
- Roundabout and office building in Veldhoven, urban development study with architectural adaptation, 1998
- Redevelopment of the centre of Sliedrecht, urban development study, 1998–1999
- Two apartment towers and a renovation project in Rotterdam, study, 1998
- 51 Patio houses on the MARIN terrain in Wageningen, study, 1998–1999